Potentiometry and Ion Selective Electrodes

Wiley India Pvt. Ltd.

Analytical Chemistry by Open Learning

Titles in Series:

1. Samples and Standards
2. Sample Pretreatment and Separation
3. Classical Methods Vol. 1
4. Classical Methods Vol. 2
5. Measurement, Statistics and Computation
6. Using Literature
7. Instrumentation
8. Chromatographic Separations
9. Gas Chromatography, 2nd Edition
10. High Performance Liquid Chromatography, 2nd Edition
11. Electrophoresis
12. Thin Layer Chromatography
13. Ultraviolet and Visible Spectroscopy
14. Fluorescence and Phosphorescence Spectroscopy
15. Biological Applications of Infra Red Spectroscopy
16. Atomic Absorption and Plasma Spectroscopy, 2nd Edition
17. Nuclear Magnetic Resonance Spectroscopy
18. X-Ray Methods
19. Mass Spectrometry, 2nd Edition
20. Scanning Electron Microscopy and X-Ray Microanalysis
21. Principles of Electroanalytical Methods
22. Potentiometry and Ion Selective Electrodes
23. Polarography and Other Voltammetric Methods
24. Radiochemical Methods
25. Clinical Specimens
26. Diagnostic Enzymology
27. Quantitative Bioassay
28. Assessment and Control of Biochemical Methods
29. Thermal Methods
30. Microprocessor Applications
31. Environmental Analysis
32. Quality in the Analytical Chemistry Laboratory

Potentiometry and Ion Selective Electrodes

Analytical Chemistry by Open Learning

Author:
ALUN EVANS
Chelmsford College of Further Education

Editor:
ARTHUR M. JAMES

on behalf of ACOL

Published on behalf of ACOL, Thames Polytechnic, London
by
JOHN WILEY & SONS
Chichester • New York • Brisbane • Toronto • Singapore

Potentiometry and Ion Selective Electrodes

Published by permission of the Controller of
Her Majesty's Stationery Office

Authorized reprint by Wiley India Pvt. Ltd., 4435/7, Ansari Road, Daryaganj,
New Delhi – 110002.

Reprint: 2008

Printed at: *Ar Emm International, New Delhi*

ISBN : 978-81-265-1732-9

Analytical Chemistry

This series of texts is a result of an initiative by the Committee of Heads of Polytechnic Chemistry Departments in the United Kingdom. A project team based at Thames Polytechnic using funds available from the Manpower Services Commission 'Open Tech' Project has organised and managed the development of the material suitable for use by 'Distance Learners'. The contents of the various units have been identified, planned and written almost exclusively by groups of polytechnic staff, who are both expert in the subject area and are currently teaching in analytical chemistry.

The texts are for those interested in the basics of analytical chemistry and instrumental techniques who wish to study in a more flexible way than traditional institute attendance or to augment such attendance. A series of these units may be used by those undertaking courses leading to BTEC (levels IV and V), Royal Society of Chemistry (Certificates of Applied Chemistry) or other qualifications. The level is thus that of Senior Technician.

It is emphasised however that whilst the theoretical aspects of analytical chemistry can be studied in this way there is no substitute for the laboratory to learn the associated practical skills. In the U.K. there are nominated Polytechnics, Colleges and other Institutions who offer tutorial and practical support to achieve the practical objectives identified within each text. It is expected that many institutions worldwide will also provide such support.

The project will continue at Thames Polytechnic to support these 'Open Learning Texts', to continually refresh and update the material and to extend its coverage.

Further information about nominated support centres, the material or open learning techniques may be obtained from the project office at Thames Polytechnic, ACOL, Wellington St., Woolwich, London, SE18 6PF.

How to Use an Open Learning Text

Open learning texts are designed as a convenient and flexible way of studying for people who, for a variety of reasons cannot use conventional education courses. You will learn from this text the principles of one subject in Analytical Chemistry, but only by putting this knowledge into practice, under professional supervision, will you gain a full understanding of the analytical techniques described.

To achieve the full benefit from an open learning text you need to plan your place and time of study.

- Find the most suitable place to study where you can work without disturbance.

- If you have a tutor supervising your study discuss with him, or her, the date by which you should have completed this text.

- Some people study perfectly well in irregular bursts, however most students find that setting aside a certain number of hours each day is the most satisfactory method. It is for you to decide which pattern of study suits you best.

- If you decide to study for several hours at once, take short breaks of five or ten minutes every half hour or so. You will find that this method maintains a higher overall level of concentration.

Before you begin a detailed reading of the text, familiarise yourself with the general layout of the material. Have a look at the course contents list at the front of the book and flip through the pages to get a general impression of the way the subject is dealt with. You will find that there is space on the pages to make comments alongside the text as you study—your own notes for highlighting points that you feel are particularly important. Indicate in the margin the points you would like to discuss further with a tutor or fellow student. When you come to revise, these personal study notes will be very useful.

Π When you find a paragraph in the text marked with a symbol such as is shown here, this is where you get involved. At this point you are directed to do things: draw graphs, answer questions, perform calculations, etc. Do make an attempt at these activities. If necessary cover the succeeding response with a piece of paper until you are ready to read on. This is an opportunity for you to learn by participating in the subject and although the text continues by discussing your response, there is no better way to learn than by working things out for yourself.

We have introduced self assessment questions (SAQ) at appropriate places in the text. These SAQs provide for you a way of finding out if you understand what you have just been studying. There is space on the page for your answer and for any comments you want to add after reading the author's response. You will find the author's response to each SAQ at the end of the text. Compare what you have written with the response provided and read the discussion and advice.

At intervals in the text you will find a Summary and List of Objectives. The Summary will emphasise the important points covered by the material you have just read and the Objectives will give you a checklist of tasks you should then be able to achieve.

You can revise the Unit, perhaps for a formal examination, by re-reading the Summary and the Objectives, and by working through some of the SAQs. This should quickly alert you to areas of the text that need further study.

At the end of the book you will find for reference lists of commonly used scientific symbols and values, units of measurement and also a periodic table.

Contents

Study Guide xiii

Practical Objectives xv

Bibliography. xvii

Acknowledgements xix

1. Summary of Basic Principles 1
- 1.1. Behaviour of Ions in Solution 1
- 1.2. Electrode Reactions 11
- 1.3. Liquid Junction Potentials. 31

2. General Principles of Potentiometry 36
- 2.1. Introduction 37
- 2.2. Ionic Strength Adjustors 43
- 2.3. Calibration 50
- 2.4. Selectivity 60
- 2.5. Limits of Measurement and Response Times. . . 65
- 2.6. Comparison of Analytical Methods 72
- 2.7. Reference Electrodes. 106
- 2.8. Choice of Experimental Requirements for an Analytical Procedure 117

3. Method of Operation of Ion-selective Electrodes . . . 121
- 3.1. The Glass Electrode 121
- 3.2. Solid State Membrane Electrodes 143
- 3.3. Electrodes based on Ion Exchange and Neutral Carriers 169
- 3.4. Gas Sensing Probes 189
- 3.5. Enzyme Electrodes 198
- 3.6. Ion-selective Field Effect Transistors. 207

4. Applications. 210
- 4.1. Case Study 1. Estimation of F ions Added to Drinking Water 210

4.2. Case Study 2. Estimation of Ca^{2+} Concentration in Beer 216
4.3. Case Study 3. Estimation of NO_3^- in Plant Tissue 219
4.4. Case Study 4. Estimation of Na^+ or Cl^- in Boiler Feed Water 221
4.5. Case Study 5. Estimation of L-tyrosine 225

Self Assessment Questions and Responses 228

Units of Measurement 299

Study Guide

This Unit is designed to give a good background to the use of potentiometry in chemical analysis. For the majority of chemists and, in fact, many non-chemists, potentiometry is one of the first instrumental analytical procedures they encounter. The use of glass electrodes in pH measurement is the most common application of potentiometry. It is used by specialists and non-specialists in applications as diverse as soil testing by gardeners and the examination of complex processes by biochemists.

The aim of this Unit is to provide sufficient knowledge for a non-specialist to select a suitable potentiometric method of analysis to solve a given problem. The Unit is designed for those wishing to extend their knowledge of the applied nature of potentiometry. The background theory included is that necessary for the understanding of what is happening. It is by no means a comprehensive text on electrochemistry. Whilst a good background in electrochemistry is very useful, it is not essential for the understanding of potentiometry. For those who are unsure of the basic principles, the first part of this Unit is a short section which summarises, very briefly, the main ideas needed in the later sections. If you have not studied electrochemistry before, you are recommended to read *ACOL: Principles of Electroanalytical Methods.*

The main body of the Unit is divided into three further Parts. The first examines the principles of the technique of potentiometry and how we can manipulate the data to give meaningful answers. We will examine the different experimental set-ups and methods used. This part of the Unit contains many worked examples and problems for you to solve. Even if you use potentiometry routinely, you may encounter some different methods and expand your existing knowledge.

Whilst potentiometry has been with us for many years, it is advances in the field of ion-selective electrodes that makes it such a valuable technique in the modern laboratory. The third part of the unit considers how the different types of electrode are constructed and relates this to the problems encountered when they are used, eg interfering ions. It is important to note that whilst we retain the term ion-selective electrode, their use is not restricted to charged ions. We can use them for a wide range of species including anions, cations and neutral species as complex as glucose and penicillin.

The final part of the Unit considers some case studies of analytical problems that have been solved using potentiometry. This is necessarily a very short selection as the applications of potentiometry would fill many volumes. It does, however, consider diverse examples.

If you are using potentiometry in your laboratory, this Unit should provide a good reinforcement of what you are doing and give you some new insights into the technique. If you are not already working in the field, the Unit can give you a good idea of the versatility of the technique.

When studying Section 2.6, although not essential, you may find it helpful to have available some Gran's plot paper (Orion Research Inc.).

Practical Objectives

Having considered the theory behind the construction and use of ion-selective electrodes, it is beneficial to carry out some practical experiments. By the end of the practical course, you should be able to:

- assemble and maintain common ion-selective and reference electrodes;
- set up a galvanic cell suitable for potentiometric analysis;
- prepare concentration and activity standards;
- calibrate ion-selective electrodes using one, two or more standards;
- perform potentiometric titrations and use the methods described in Section 2.6.3. to obtain the equivalence point;
- analyse a number of similar samples using a calibration graph;
- perform analyses using addition and multiple addition procedures;
- devise a method for analysing a sample containing an interfering ion.

Bibliography

The following texts deal exclusively with potentiometry and ion selective electrodes.

– P. L. Bailey *Analysis with ion-selective Electrodes* (2nd Edn), Heyden, London, 1980.

– H. Freiser, *Ion-Selective Electrodes in Analytical Chemistry*– Vols. 1 and 2, Plenum, New York, 1980.

The following are general analytical chemistry text books which have good chapters on potentiometry. They will not be as detailed as the above.

– G. Christian, *Analytical Chemistry*, Wiley, New York, 1980.

– J. S. Fritz & G. H. Schenk, *Quantitative Analytical Chemistry* (4th Edn), Allyn & Bacon, Boston, 1979.

– S. E. Manahan, *Quantitative Chemical Analysis*, Brooks/Cole, Monerey Calif., 1986.

Information produced by electrode manufacturers is also very helpful. The *Analytical Method Guide* produced by Orion Research is a particularly good example of the type of material available (often free) from such sources.

Acknowledgements

Figures 2.5b, 3.3a and 3.3b are redrawn from Orion Research instruction manuals. Permission applied for.

Figure 3.4a is redrawn from the EIL ammonia probe manual with permission from Kent Industrial Measurements Ltd.

The author wishes to acknowledge the great debt he owes to the late Dr H. E. Hallam of University College, Swansea for his support and inspiration as both an undergraduate tutor and research supervisor.

1. Summary of Basic Principles

Overview

This first Part of the Unit on potentiometry is a basic summary of the electrochemical principles you will need to know when following through this Unit. It is not intended as a comprehensive treatment, rather as a revision of some important areas. If you have not studied electrochemistry before, you are recommended to read the introductory Unit in this series.

In this Part you will cover the behaviour of ions in solution and examine how the behaviour of a given ion is modified by other ions. You will then examine electrode processes and the idea of galvanic cells. Finally, you will see how changes in the activity of a given ion affects the emf of a galvanic cell and how we can use this in quantitative analysis.

1.1. BEHAVIOUR OF IONS IN SOLUTION

Whilst some electroanalytical techniques can be used to monitor molecules having no net charge, the majority rely on the species under investigation having a net charge or gaining or losing charge in a redox reaction. In potentiometry we generally encounter charged species or ions. Solutions containing ions conduct electricity and are termed electrolytes and these may be subdivided into true electrolytes and potential electrolytes.

True electrolytes are ionic compounds, such as sodium chloride, which on dissolving in water undergo a transformation which can be described by the following equation:

$$Na^+Cl^-(s) + H_2O(l) \rightarrow Na^+(aq) + Cl^-(aq)$$

where (s) implies that the substances is in the solid state and it will consist of positively charged sodium ions and negatively charged chloride ions arranged into a regular pattern or crystal. Each sodium ion will be surrounded by six chloride ions and vice-versa. Similarly, (l) refers to water in its liquid state and (aq) to a solvated ion. This means that each ion is surrounded by a number of water molecules, which effectively insulate it from the other ions. The water molecule is polar. Polar molecules have no overall charge, but the electrons in the molecule are displaced towards one of the atoms, in this case the oxygen. This gives a small negative charge on the oxygen and a small positive charge on the hydrogen. Thus, water is able to surround both positive and negative ions as shown in Fig. 1.1a.

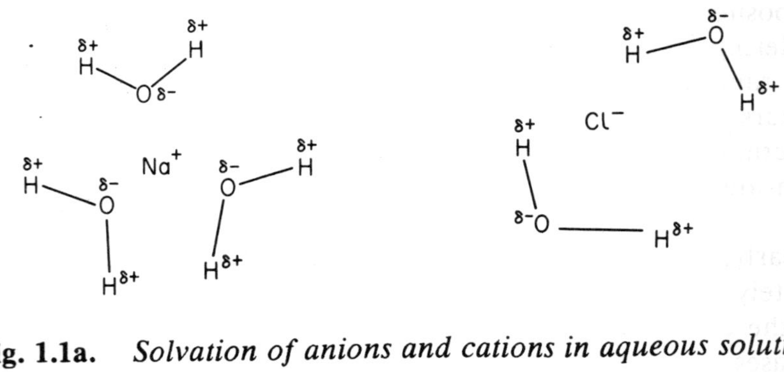

Fig. 1.1a. *Solvation of anions and cations in aqueous solution*

Potential electrolytes, on the other hand, are covalent molecules which react with water to form ions. An example of such an electrolyte is gaseous hydrogen chloride which reacts

$$HCl(g) + H_2O(l) \rightarrow H_3O^+(aq) + Cl^-(aq)$$

where (g) and (l) refer to gas and liquid states respectively and, as with true electrolytes, (aq) refers to a solvated ion. It should be noted, however, that the behaviour of solvated ions is independent of the source and the $Cl^-(aq)$ from NaCl is identical to that from HCl.

In very dilute solutions, there will be sufficient water molecules to fully surround each ion and this circumstance is referred to as complete solvation. As the concentration is low, the distance between a given ion and its nearest neighbour will be relatively large. Each ion can then be considered as a separate entity which will behave independently of all other ions present. This situation is referred to as ideal behaviour.

However, at the other extreme, with very concentrated solutions the situation changes. Firstly, there may be insufficient water molecules to completely solvate each ion. This means that an ion may not be fully 'insulated' from its neighbours. Secondly, the ions will be very much closer together and forces of attraction and repulsion between opposite and like charged ions respectively will occur. Thus, each individual ion's behaviour will be influenced by the other ions present. An ion of one charge will be surrounded by an ion atmosphere of oppositely charged ions. Whilst this gives rise to some short term ordering within the solution, it is not a permanent arrangement in the same way as a solid crystal. Random motion of solvated ions occurs and the composition of the ion atmosphere changes. Such solutions have slightly modified properties and are referred to as non-ideal.

Clearly, ideal behaviour only exists when the concentration is infinitely small (the expression infinite dilution is often used) and as the concentration increases, the degree of non-ideality also increases. As ions are surrounded by oppositely charged ions, any electrochemical measurements are affected and the solutions behave as if their concentration were less. This leads to use of a function termed activity (a), which can crudely be defined as the thermodynamic or effective concentration. The ion atmosphere, therefore, effectively reduces the concentration by a factor termed the activity coefficient (γ). If the concentration is represented by c, then

$$a = \gamma c \tag{1.1a}$$

If the activity coefficient is unity, then $a = c$ and the solution behaves ideally. As the degree of non-ideality increases (ie as the concentration and the effect of other ions increases) the activity coefficient will become smaller.

Debye and Hückel carried out a theoretical treatment of ionic solutions in an attempt to predict values of the activity coefficient and, hence, activity. They identified two major factors which influenced the value of the activity coefficient:

- the proximity of each ion to its nearest neighbour, which is related to the concentration. This is because the force of attraction between oppositely charged particles obeys an inverse square law;

- the charges on the ions. The force of attraction between doubly charged ions (such as Cu^{2+} and SO_4^{2-}) will be greater than between singly charged ions (such as Na^+ and Cl^-).

To allow for these factors, they defined a term, total ionic strength (I) as:

$$I = 0.5 \sum_i c_i z_i^2 \tag{1.1b}$$

where c_i and z_i are the stoichiometric concentration and charge respectively on a given ion, i. The symbol, $\sum$ (Greek sigma), refers to the sum of all such terms as $c_i z_i^2$. The total ionic strength of a solution is, therefore, calculated by multiplying the concentration of each individual ion by the square of its charge and then adding all of these values before dividing by two.

Π Calculate the total ionic strength of a 0.100 mol dm^{-3} solution of copper(II)sulphate(VI).

Copper(II)sulphate(VI) dissolves

$$CuSO_4(s) + H_2O(l) \rightarrow Cu^{2+}(aq) + SO_4^{2-}(aq)$$

and every mole of $CuSO_4$ produces one mole of Cu^{2+} ions and one mole of SO_4^{2-} ions. Thus

$$[Cu^{2+}] = 0.100 \text{ mol dm}^{-3}$$

$$[SO_4^{2-}] = 0.100 \text{ mol dm}^{-3}$$

The total ionic strength is then calculated:

$$I = 0.5([Cu^{2+}]2^2 + [SO_4^{2-}]2^2)$$

$$= 0.5([0.100]4 + [0.100]4)$$

$$= 0.400 \text{ mol dm}^{-3}$$

Notice how the concentration of each ion is multiplied by the square of the charge before addition and dividing by two.

SAQ 1.1a

Calculate the total ionic strength of a 0.100 mol dm^{-3} solution of sodium sulphate(VI) (note that the formula of sodium sulphate(VI) is Na_2SO_4 and there will be twice as many sodium ions as sulphate(VI) ions).

SAQ 1.1a

The above examples deal with solutions of one substance giving rise to two different ions. However, once dissolved, the ion does not remember its origins and one sulphate ion is very much like another. This allows us to calculate the total ionic strength of solutions prepared from two or more substances. This is best illustrated by the following example.

Π Calculate the total ionic strength of a solution prepared by dissolving 0.100 mole of sodium chloride and 0.200 mole of sodium sulphate(VI) in one dm^3 of water.

The sodium chloride dissolves:

$$NaCl(s) + H_2O(l) \rightarrow Na^+(aq) + Cl^-(aq)$$

and the sodium sulphate:

$$Na_2SO_4(s) + H_2O(l) \rightarrow 2\,Na^+(aq) + SO_4^{2-}(aq)$$

The total concentration of sodium ions must reflect both sources:

$$[Na^+] = 0.100 + (2 \times 0.200) \text{ mol dm}^{-3}$$

$$= 0.500 \text{ mol dm}^{-3}$$

The remaining two ions are derived from one source each:

$$[Cl^-] = 0.100 \text{ mol dm}^{-3}$$

$$[SO_4^{2-}] = 0.200 \text{ mol dm}^{-3}$$

The total ionic strength now involves three ions:

$$\begin{aligned} I &= 0.5([Na^+]1^2 + [Cl^-]1^2 + [SO_4^{2-}]2^2) \\ &= 0.5([0.500]1 + [0.100]1 + [0.200]4) \\ &= 0.700 \text{ mol dm}^{-3} \end{aligned}$$

The process can be extended to give the total ionic strength of any solution with any number of components. The total ionic strength is a property of a given solution and not of a given ion. In practice, we are concerned with the behaviour of a given ion and not the solution as such. However, the total ionic strength plays a large part in determining the extent of the ion atmosphere as it reflects both the concentration and charge of all the ions present. Debye and Hückel went on to show that by considering the electrostatic forces present in solution the activity coefficient of a given ion is related to the total ionic strength by the expression:

$$\log \gamma_i = -0.5091 z_i^2 I^{\frac{1}{2}} \qquad (1.1c)$$

in aqueous solution at 298 K. This expression allows the activity coefficient and, consequently, the activity of any ion to be calculated for solutions of concentration below 10^{-3} for 1 : 1 electrolytes and below 10^{-4} mol dm^{-3} for 1 : 2 electrolytes. These limits lead to the expression being termed the Debye–Hückel limiting law.

Π Calculate the activity of the chloride ion in each of the following solutions:

(*i*) 0.100 mol dm^{-3} sodium chloride;

(*ii*) a solution prepared by dissolving 0.100 mole of sodium chloride and 0.200 mole of sodium sulphate(VI) in 1.000 dm^3 of water.

For the first solution:

$$I = 0.5([0.100 \times 1^2] + [0.100 \times 1^2])$$

$$= 0.100 \text{ mol dm}^{-3}$$

The activity coefficient is given by:

$$\log \gamma(\text{Cl}^-) = -0.5091(1)^2 (0.100)^{\frac{1}{2}}$$

$$= -0.161$$

$$\gamma(\text{Cl}^-) = \text{antilog}(-0.161)$$

$$= 0.690$$

the activity is then given by:

$$a(\text{Cl}^-) = 0.690 \times 0.100$$

$$= 0.0690 \text{ mol dm}^{-3}$$

The second solution is the same as that in the previous question. The total ionic strength is 0.700 mol dm^{-3}. The activity coefficient is calculated as follows:

$$\log\gamma(\text{Cl}^-) = -0.5091(1)^2 (0.700)^{\frac{1}{2}}$$

$$= -0.426$$

$$\gamma(\text{Cl}^-) = \text{antilog}(-0.426)$$

$$= 0.375$$

The activity is then given by:

$$a(\text{Cl}^-) = 0.375 \times 0.100$$

$$= 0.0375 \text{ mol dm}^{-3}$$

Clearly, the activity of the chloride ion is much less in the second example than in the first. This is due to the effect of the additional ions provided by the sodium sulphate(VI).

In electrochemistry, therefore, it is important to relate the measured properties to the activities of the ions and not to the concentrations.

Summary

In this section we have examined briefly how the behaviour of one ion is affected by the presence of other ions in solution. In electroanalytical chemistry it is, therefore, important to relate any measurement to the activity of the ion, rather than to its concentration.

SAQ 1.1b

State the two factors which influence the extent of the ion atmosphere surrounding a given ion.

SAQ 1.1c

What term is used to reflect the 'effective concentration' of a given ion?

SAQ 1.1d

Calculate the activity of the copper ion in a 0.250 mol dm^{-3} solution of copper(II) chloride($CuCl_2$).

Objectives

You should now be able to:

- list the two factors which influence the extent of the ion atmosphere surrounding a given ion;
- define the terms activity, activity coefficient and total ionic strength;
- calculate the activity of anion given the concentrations of the ionic compounds present in the solution.

1.2. ELECTRODE REACTIONS

Overview

In this section we will consider the redox reactions that occur at electrodes. We will then extend the picture to the formation of galvanic cells and the relationship between the activity of ions in solution and the measured emf. It is this latter relationship which is the basis of the analytical technique of potentiometry.

1.2.1. Redox Reactions at Electrodes

Electrode reactions involve the transfer of charge from the electrode to the dissolved species or vice versa. Reactions involving this transfer of charge from one species to another are termed Redox Reactions. The name redox is a composite of REDuction and OXidation.

An oxidation reaction is one in which a species loses an electron (negative charge). For example, copper can lose two electrons to form a copper(II) ion

$$Cu(s) \rightarrow Cu^{2+}(aq) + 2e$$

The copper is said to be oxidised.

A reduction reaction is one in which a species gains an electron. This process is the reverse of oxidation and the copper(II) ion can be reduced to copper

$$Cu^{2+}(aq) + 2e \rightarrow Cu(s)$$

ie oxidation and reduction can be summarised by:

$$\text{(oxidised state)} + ne \rightleftharpoons \text{(reduced state)}$$

which is the general equation for all electrode reactions. In practice an oxidation reaction cannot take place without a corresponding reduction and vice versa. For example, when zinc metal is dropped into a solution of copper(II) ions a red-brown precipitate of copper soon appears on the surface of the zinc. The overall reaction is written as

$$Zn(s) + Cu^{2+}(aq) \rightarrow Cu(s) + Zn^{2+}(aq)$$

This overall reaction can be split into the appropriate oxidation and reduction processes:

$$Zn(s) \rightarrow Zn^{2+}(aq) + 2e \qquad \text{oxidation}$$

$$Cu^{2+}(aq) + 2e \rightarrow Cu(s) \qquad \text{reduction}$$

The two separate ionic equations are sometimes referred to as half-equations. Note that the number of electrons lost by the zinc equals the number of electrons gained by the copper.

Π Classify the following into oxidation or reduction processes:

(*a*) $Ba^{2+} + 2e \rightarrow Ba(s)$

(*b*) $Cl_2(g) + 2e \rightarrow 2\,Cl^-(aq)$

(*c*) $Fe^{2+}(aq) \rightarrow Fe^{3+}(aq) + e$

(*d*) $MnO_4^-(aq) + 8\,H^+(aq) + 5e \rightarrow Mn^{2+}(aq) + 4\,H_2O(l)$

Reaction (*a*) is a reduction reaction as it involves electron gain.

Reaction (*b*) is also a reduction reaction as it involves electron gain.

Reaction (*c*) is an oxidation reaction as it involves the loss of an electron.

Reaction (*d*) is a reduction reaction as it involves electron gain.

SAQ 1.2a

Write down the half-equations corresponding to the oxidation and reduction processes for the following overall reactions:

(*i*) $Fe^{2+}(aq) + Ce^{4+}(aq) \rightarrow Fe^{3+}(aq) + Ce^{3+}(aq)$

(*ii*) $Zn(s) + 2\,Ag^+(aq) \rightarrow Zn^{2+}(aq) + 2\,Ag(s)$

(*iii*) $Fe(s) + S(s) \rightarrow FeS(s)$

SAQ. 1.2a

1.2.2. Galvanic Cells

Electrode processes are redox reactions that occur at the interface of a metal or other solid conductor (the electrode) and a solution. The electrode itself may or may not be directly involved in this redox reaction. For example, if a rod of copper metal dips into a solution of copper(II) ions, then two processes are possible. Firstly, the copper may be oxidised and dissolve in the solution as copper(II) ions:

$$Cu(s) \rightarrow Cu^{2+}(aq) + 2e$$

Alternatively, the copper ions may be oxidised and 'plate out' on the electrode as metallic copper:

$$Cu^{2+}(aq) + 2e \rightarrow Cu(s)$$

In each of these two processes, the electrode is chemically involved in the redox reaction. Any net change in the number of free electrons is accommodated by the metallic bond existing in the electrode. If oxidation occurs, the positively charged copper ions dissolve in the solution leaving the electrons delocalised throughout the metallic lattice. In this way the solution becomes positively charged with respect to the electrode. In the reverse process, the copper ions in the solution will take electrons from the electrode before depositing on the surface. The electrode is the deficient of electrons and it will now be positively charged with respect to the solution.

An example of an electrode reaction in which the electrode itself remains chemically unchanged involves platinum metal dipping into a mixture of iron(II) and iron(III) ions. If the iron(II) is oxidised:

$$Fe^{2+}(aq) \rightarrow Fe^{3+}(aq) + e$$

the electron released is accepted by the electrode and delocalised through the metallic lattice of the platinum. The electrode then has a net negative charge relative to the solution. If, on the other hand, the iron(III) is reduced:

$$Fe^{3+}(aq) + e \rightarrow Fe^{2+}(aq)$$

then an electron leaves the platinum lattice to perform the oxidation. This leaves the electrode with a positive charge relative to the solution.

In each of these two processes, there is a tendency for a charge difference to exist between the electrode and the solution. This is illustrated in a Daniell cell illustrated in Fig. 1.2a. The cell consists of a copper electrode dipping into a solution of copper(II) ions and a zinc electrode dipping into a solution of zinc(II) ions. Electrical conductance between the two solutions is achieved using an inverted tube containing a solution of potassium chloride. If a metal wire was used, then the picture would be complicated due to the introduction of two more electrodes into the circuit. The zinc and copper electrodes are then connected by an external circuit which can be a high impedance voltmeter or other suitable device to indicate a potential difference between the two.

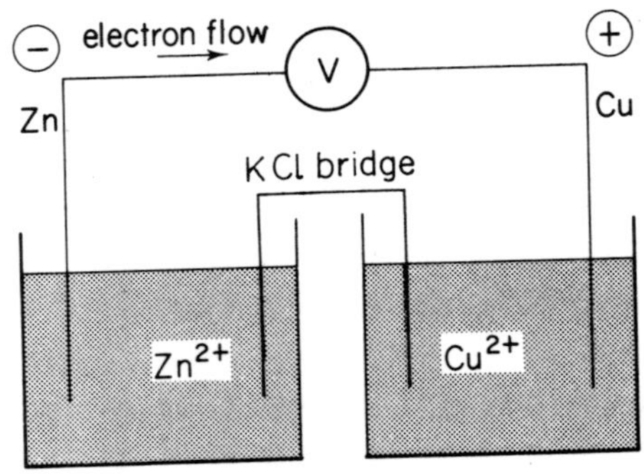

Fig 1.2a. *Diagrammatic representation of a Daniell cell*

The Daniell cell can, therefore be described in terms of two half cells, where a half cell is the combination of an electrode and the solution with which it is in contact. One half cell is due to Cu^{2+}/Cu and the tendency is for this system to lie towards the reduced form. That is, the tendency is for the copper(II) ions to be reduced to copper:

$$Cu^{2+}(aq) + 2e \rightarrow Cu(s)$$

The electrode is, therefore, positively charged relative to the solution. On the other hand, the tendency in the zinc half cell lies in the opposite direction. The zinc is likely to be oxidised into zinc(II) ions:

$$Zn(s) \rightarrow Zn^{2+}(aq) + 2e$$

The electrode will then be negatively charged relative to the solution.

As the solutions are connected by the KCl bridge, their potentials can be assumed to be equal and, consequently, we have a system which can be represented by an electrical circuit as shown below:

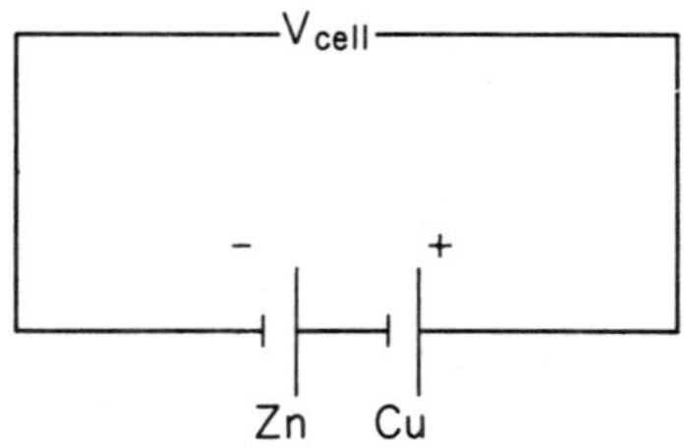

The voltmeter will then measure the difference in potential between the two electrodes. In fact the situation is identical to connecting up two batteries in series.

Note that the one half cell involves an oxidation process and the other half cell a reduction process. These are then combined in the cell to give a redox change represented by:

$$Zn(s) + Cu^{2+}(aq) \rightarrow Zn^{2+}(aq) + Cu(s)$$

When the cell is in use, the zinc electrode will gradually dissolve, whilst the copper electrode will gradually increase in size.

This concept of combining two different half cells to produce electricity can be extended to a number of systems. The general name for cells of this type is galvanic cell and to simplify their representation, use is made of cell notation or cell nomenclature. This is a shorthand way of writing down cells and is best illustrated with an example.

Consider the Daniell Cell illustrated in Fig. 1.2a. The IUPAC convention for writing down this cell is

$$\ominus\ Zn \mid Zn^{2+} \parallel Cu^{2+} \mid Cu\ \oplus$$

This notation starts with the left hand electrode and moves to the right through the solutions to the right hand electrode. The single vertical bars signify a phase boundary, whilst the double vertical bar the KCl bridge. This convention can be extended to include the activities of the two ions. For example

$$\ominus\ Zn|Zn^{2+}\ (a = 1.0\ \text{mol dm}^{-3})||Cu^{2+}\ (a = 1.0\ \text{mol dm}^{-3})|Cu\ \oplus$$

SAQ 1.2b

Draw the galvanic cell represented by the following cell notation.

Pt | Fe^{3+},Fe^{2+} || Cu^{2+} | Cu

1.2.3. Electrode Potentials

We have seen that a potential difference can be established between an electrode and a solution. We now need a system which allows us to estimate the magnitude and sign of this potential. The potential difference between the electrode and solution in a half cell is referred to as the electrode potential and for comparison purposes, all electrode potentials are written for the reduction process, ie for the a metal, M and its own ion, M^{n+}, it is written as

$$M^{n+}(aq) + ne \rightleftharpoons M(s)$$

The electrode potential is, therefore, often referred to as the reduction potential. In a galvanic cell, one half cell exhibits the reduction process, whilst the other shows an oxidation process. Cell convention states that oxidation is taken to occur at the left hand electrode, which then releases electrons to the external circuit. This electrode is referred to as the anode, ie *oxidation always occurs at the anode*.

The reduction process is taken to occur at the right hand electrode and this is referred to as the cathode. Note that *reduction always occurs at the cathode*.

Calculating cell potentials is the same as calculating the potential of two batteries in series. As the left hand half cell undergoes oxidation, its electrode potential must be subtracted from that of the right hand half-cell. The emf of the cell is then given by

$$E(\text{cell}) = E(\text{right}) - E(\text{left})$$

where E(cell) is the emf of the cell and E(right) and E(left) are the electrode (reduction) potentials of the right and left half cells respectively. Remember that the negative sign compensates for the electrode reaction occurring in the reverse direction. For the Daniell cell, the cell emf is given by

$$E(\text{cell}) = E(Cu^{2+}/Cu) - E(Zn^{2+}/Zn)$$

The way that the chemical equation representing the redox change that occurs in the cell must be consistent with this cell convention. If the oxidation process occurs at the left hand electrode, the equation is written so that the electrons are given up. For the Daniell cell in Fig. 1.2a, the left hand half cell is the zinc and:

$$Zn(s) \rightarrow Zn^{2+}(aq) + 2e$$

Conversely, the right hand half cell reaction is written as a reduction and in the Daniell cell, this will be the copper:

$$Cu^{2+}(aq) + 2e \rightarrow Cu(s)$$

Combining these two equations gives the overall cell reaction:

$$Zn(s) + Cu^{2+}(aq) + 2e \rightarrow Zn^{2+}(aq) + 2e + Cu(s)$$

The electrons cancel out to give:

$$Zn(s) + Cu^{2+}(aq) \rightarrow Zn^{2+}(aq) + Cu(s)$$

Π Write down the overall cell equation for the galvanic cell represented by

$$Pt \mid Fe^{3+}, Fe^{2+} \parallel Cu^{2+} \mid Cu$$

Oxidation occurs at the left hand electrode, ie the iron(II) will be oxidised. The chemical equation is then:

$$Fe^{2+}(aq) \rightarrow Fe^{3+}(aq) + e$$

Reduction occurs at the right hand electrode, which accepts electrons from the external circuit. The chemical equation is then written:

$$Cu^{2+}(aq) + 2e \rightarrow Cu(s)$$

To give the overall cell equation the number of electrons must balance. To achieve this the iron(II)/iron(III) equation is multiplied by two and then added to the copper/copper(II) equation:

$$2\,Fe^{2+}(aq) + Cu^{2+}(aq) + 2e \rightarrow 2\,Fe^{3+}(aq) + 2e + Cu(s)$$

cancelling the electrons gives:

$$2\,Fe^{2+}(aq) + Cu^{2+}(aq) \rightarrow 2\,Fe^{3+}(aq) + Cu(s)$$

Having established this convention for writing the equations corresponding to cell notation, it is possible to make use of some experimental values of electrode potentials.

1.2.4. Standard Electrode Potentials

In order that valid comparisons are made of the magnitudes of the electrode potentials, a set of standard conditions are adopted:

- all measurements are made at 298 K;
- all solutes present are at unit activity (ie one mol dm^{-3});
- all half cells are measured relative to a standard half cell termed the standard hydrogen electrode, abbreviated to she.

The electrode potential measured under these conditions is termed a standard electrode potential or sep. It is denoted by the symbol $E^{\ominus}$. The actual procedure for determining values for sep is quite involved and does not concern us here. We will only concern ourselves with their use under the conditions encountered in potentiometry.

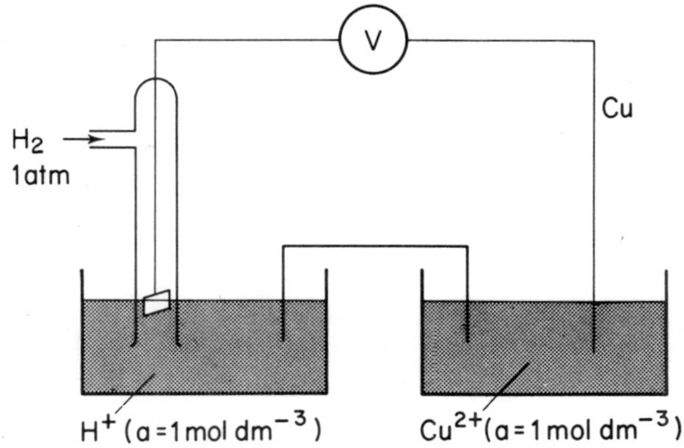

Fig. 1.2b. *Diagrammatic representation of a galvanic cell for the measurement of the sep of the copper(II)/copper half cell*

An sep is measured as a reduction process. The half cell under investigation is therefore, represented on the right hand side of the cell. The she then forms the left hand half cell. The cell for the determination of the sep of the copper(II)/copper half cell is illustrated in Fig. 1.2b and is represented by the following cell notation:

$$Pt \mid H_2(1\ atm) \mid H^+(a = 1\ mol\ dm^{-3}) \| Cu^{2+}(a = 1\ mol\ dm^{-3}) \mid Cu$$

The two half cell equations are:

$$H_2(g) \rightarrow 2\,H^+(aq) + 2e \qquad \text{left/oxidation}$$

$$Cu^{2+}(aq) + 2e \rightarrow Cu(s) \qquad \text{right/reduction}$$

Combining these two equations and cancelling out the electrons gives an overall cell equation of:

$$Cu^{2+}(aq) + H_2(g) \rightarrow Cu(s) + 2\,H^+(aq)$$

The emf of this cell is given by

$$E(\text{cell}) = E(\text{right}) - E(\text{left})$$

or, since each half cell is in its standard state

$$E^{\ominus}(\text{cell}) = E^{\ominus}(Cu^{2+}/Cu) - E^{\ominus}(\text{she})$$

The value for the sep of the she is taken as zero volts by convention. This is an arbitrary decision which allows us a point of reference to measure the electrode potential of other half cells against.

An anology is the use of 'sea level' as a reference point of zero height. Mountains then have positive heights relative to this standard, whilst the sea bed will have negative values.

Thus as

$$E^{\ominus}(\text{she}) = 0\ V$$

$$E^{\ominus}(\text{cell}) = E^{\ominus}(Cu^{2+}/Cu)$$

Thus, the cell illustrated will measure the sep of the copper(II)/copper half cell directly. The standard electrode potentials of other half cells may be measured by substituting the right hand half cell in Fig. 1.2b. Values of sep for a number of different systems are given in Fig. 1.2c.

SAQ 1.2c

Draw the cell for the determination of the sep of the iron(III)/iron(II) equilibrium.

Electrode	Electrode Reaction	$E^\ominus$/V
$Pt \mid F_2 \mid F^-$	$F_2 + 2e = 2F^-$	+2.87
$Pt \mid H_2O_2 \mid H^+$	$H_2O_2 + 2H^+ + 2e = 2H_2O$	+1.77
$Pt \mid MnO_4^-, Mn^{2+}$	$MnO_4^- + 8H^+ + 5e = Mn^{2+} + 4H_2O$	+1.51
$Pt \mid Cl_2 \mid Cl^-$	$Cl_2 + 2e = 2Cl^-$	+1.3595
$Pt \mid Tl^{3+}, Tl^+$	$Tl^{3+} + 2e = Tl^+$	+1.25
$Pt \mid Br_2 \mid Br^-$	$Br_2 + 2e = 2Br^-$	+1.065
$Ag^+ \mid Ag$	$Ag^+ + e = Ag$	+0.7991
$Pt \mid Fe/^{3+}, Fe^{2+}$	$Fe^{3+} + e = Fe^{2+}$	+0.771
$Pt \mid O_2 \mid H_2O_2$	$O_2 + 2H^+ + 2e = H_2O_2$	+0.682
$Pt \mid I_2 \mid I^-$	$I^{3-} + 2e = 3I^-$	+0.536
$Cu^{2+} \mid Cu$	$Cu^{2+} + 2e = Cu$	+0.337
$Pt \mid Hg_2Cl_2 \mid Hg \mid Cl^-$	$Hg_2Cl_2 + 2e = 2Cl^- + 2Hg$	+0.2676
$AgCl \mid Ag \mid Cl^-$	$AgCl + e = Ag + Cl^-$	+0.2225
$Pt \mid Cu^{2+}, Cu^+$	$Cu^{2+} + e = Cu$	+0.153
$CuCl \mid Cu \mid Cl^-$	$CuCl + e = Cu + Cl^-$	+0.137
$AgBr \mid Ag \mid Br^-$	$AgBr + e = Ag + Br^-$	+0.0713
$Pt \mid H^+ \mid H_2$	$2H^+ + 2e = H_2$	0.0000
$Pb^{2+} \mid Pb$	$Pb^{2+} + 2e = Pb$	−0.126
$AgI \mid Ag \mid I^-$	$AgI + e = Ag + I^-$	−0.1518
$CuI \mid Cu \mid I^-$	$CuI + e = Cu + I^-$	−0.1852
$PbSO_4 \mid Pb \mid SO_4^{2-}$	$PbSO_4 + 2e = Pb + SO_4^{2-}$	−0.3588
$Pt \mid Ti^{3+}, Ti^{2+}$	$Ti^{3+} + e = Ti^{2+}$	−0.369
$Cd^{2+} \mid Cd$	$Cd^{2+} + 2e = Cd$	−0.403
$Fe^{2+} \mid Fe$	$Fe^{2+} + 2e = Fe$	−40.4402
$Cr^{3+} \mid Cr$	$Cr^{3+} + 3e = Cr$	−0.744
$Zn^{2+} \mid Zn$	$Zn^{2+} + 2e = Zn$	−0.7628
$Mn^{2+} \mid Mn$	$Mn^{2+} + 2e = Mn$	−1.180
$Al^{3+} \mid Al$	$Al^{3+} + 3e = Al$	−1.662
$Mg^{2+} \mid Mg$	$Mg^{2+} + 2e = Mg$	−42.363
$Na^+ \mid Na$	$Na^+ + e = Na$	−2.7142
$Ca^{2+} \mid Ca$	$Ca^{2+} + 2e = Ca$	−2.866
$Ba^{2+} \mid Ba$	$Ba^{2+} + 2e = Ba$	−2.906
$K^+ \mid K$	$K^+ + e = K$	−2.9252
$Li^+ \mid Li$	$Li^+ + e = Li$	−3.045

Fig. 1.2c. *Standard electrode potentials at 298 K*

The values quoted in Fig. 1.2c can now be used to calculate the emf of galvanic cells using the rules already established.

Π Calculate the emf of the Daniell Cell represented by:

$$Zn \mid Zn^{2+}(a = 1 \text{ mol dm}^{-3}) \parallel Cu^{2+}(a = 1 \text{ mol dm}^{-3}) \mid Cu$$

Applying:

$$\begin{aligned} E(\text{cell}) &= E(\text{right}) - E(\text{left}) \\ &= E^{\ominus}(\text{Cu}) - E^{\ominus}(\text{Zn}) \\ &= 0.337 - (-0.7628) \\ &= 1.100 \text{ V} \end{aligned}$$

This emf applies to the cell described by the above cell notation and will apply to the chemical equation

$$Zn + Cu^{2+} \rightarrow Zn^{2+} + Cu$$

Let us now consider the effect of writing the Daniell cell notation in the reverse direction.

Π Using the cell convention described earlier write down the chemical equation and calculate the cell emf of the cell represented by:

$$Cu \mid Cu^{2+}(a = 1 \text{ mol dm}^{-3}) \parallel Zn^{2+}(a = 1 \text{ mol dm}^{-3}) \mid Zn$$

Left Hand Electrode $Cu \rightarrow Cu^{2+} + 2e$

Right Hand Electrode $Zn^{2+} + 2e \rightarrow Zn$

Overall Equation $Cu + Zn^{2+} \rightarrow Cu^{2+} + Zn$

$$\begin{aligned} E(\text{cell}) &= E^{\ominus}(Zn^{2+}/Zn) - E^{\ominus}(Cu^{2+}/Cu) \\ &= -0.7628 - (+0.337) \\ &= -1.100 \text{ V} \end{aligned}$$

The answers to the above questions show that the chemical equation is written in the reverse direction, ie

$$Cu + Zn^{2+} \rightarrow Cu^{2+} + Zn$$

and the cell emf becomes a negative value at −1.100 V.

It can be seen that whilst the magnitude of $E^{\ominus}$ for the cell remains the same its sign changes. This is a very important observation. Experiments show that whilst zinc will react spontaneously with copper(II), copper will not react spontaneously with zinc(II) ions.

As a general rule, if the sign of $E^{\ominus}$ for the cell is positive, the reaction will proceed spontaneously in the direction written. If the sign of $E^{\ominus}$ for the cell is negative, the reaction will proceed in the *reverse* direction. This is consistent with the idea of free energy change $\Delta G^{\ominus}$). As

$$\Delta G^{\ominus} = - nFE^{\ominus}$$

If $E^{\ominus}$ is positive, then $\Delta G^{\ominus}$ will be negative and the reaction proceed spontaneously in the direction written.

SAQ 1.2d

Write down the overall chemical equation for the cell represented by:

$$Pt \left| \begin{matrix} Fe^{3+} & Fe^{2+} \\ (a = 1\ mol\ dm^{-3}); & (a = 1\ mol\ dm^{-3}) \end{matrix} \right\| \begin{matrix} Zn^{2+} \\ (a = 1\ mol\ dm^{-3}) \end{matrix} \Big| \; Zn$$

What is the standard emf of this cell and in which direction will the reaction proceed?

SAQ 1.2d

1.2.5. Variation of Cell emf with Activity

The standard conditions outlined in the previous section listed a fixed activity of one mol dm^{-3} for each solute present. If the activity of the oxidised or reduced species differs from unity, then the electrode potential of the half cell will change from the standard value. The Nernst equation may be used to compensate for changes in activity and this is one of the most important equations in electrochemistry. When applied to a half cell it states that:

$$E = E^{\ominus} - \frac{RT}{nF} \ln \frac{a(\text{reduced form})}{a(\text{oxidised form})} \qquad (1.2a)$$

where E is the measured emf of the half cell:

The term 'ln' refers to natural logarithm (ie to the base e). In practice, it is sometimes more convenient to use logarithm to the base 10 (written as log) because activities are generally quoted to powers of ten, eg 1.0×10^{-3} mol dm^{-3}. As

$$\ln a = 2.303 \log a$$

the Nernst equation is then written as:

$$E = E^{\ominus} - 2.303 \frac{RT}{nF} \log \frac{a(\text{reduced form})}{a(\text{oxidised form})} \qquad (1.2b)$$

For the copper(II)/copper system, the oxidised form is Cu^{2+} and the reduced form is Cu. The number of electrons in the equation is 2. The Nernst equation is then written as:

$$E = E^{\ominus} - 2.303 \frac{RT}{nF} \log \frac{a(\text{Cu})}{a(\text{Cu}^{2+})}$$

The activity of any solid is unity as it is in its standard state. Therefore,

$$E = E^{\ominus} - 2.303 \frac{RT}{nF} \log \frac{1}{a(\text{Cu}^{2+})}$$

and as $\log(1/x) = -\log x$

The Nernst equation applies to each half cell in turn. The emf of the cell under non-standard state conditions is then obtained by subtracting the E (not the $E^{\ominus}$) values obtained from the Nernst equation in the normal way.

Π Calculate the electrode potential of the copper(II)/copper half cell if the activity of the copper ions is 0.100 mol dm^{-3} at 298 K.

From Fig. 1.2c, the $E^{\ominus}$ value for copper is +0.337 V.

The Nernst equation for the copper system is written as:

$$E = E^{\ominus} - 2.303 \frac{RT}{nF} \log a(\text{Cu}^{2+})$$

substituting for $a(\text{Cu}^{2+})$ and the constants (remember that $n = 2$) we have:

$$E = 0.337 + 0.0296 \log 0.100$$

$$= 0.307 \text{ V}$$

The calculation indicates that the electrode potential of the half cell containing the copper(II) ion of activity 0.100 mol dm^{-3} is +0.307 V. The standard electrode potential is +0.337 V. Thus, a change in activity produces a change in the electrode potential of the half cell and, consequently, in the measured cell emf. This is the basis of the analytical technique of potentiometry, which can be used to measure the activity and, in turn, concentration of a given species in solution.

Some of the chemical equations representing half cells, contain more than one species on the oxidised and reduced form of the equation. The Nernst equation applied to such systems uses the product of the activities of all species on the reduced side divided by the product of the activities of the species on the oxidised side. For example, consider the half cell:

$$H_2O_2 \mid H^+ \mid Pt$$

The cell equation is:

$$H_2O_2 + 2\,H^+ + 2e \rightarrow 2\,H_2O$$

The multiple 2 in front of both the H^+ and the H_2O means that two molecules of each are involved in the reaction. Thus, the activity of these two species is squared in the Nernst equation ($a \times a = a^2$). The Nernst equation is then written as:

$$E = E^{\ominus} - 2.303\,\frac{RT}{2F}\,\log\,\frac{a^2(H_2O)}{a^2(H_2O_2)a^2(H^+)}$$

In the majority of applications, however, we are not concerned with such complex forms of the equation.

SAQ 1.2e

A galvanic cell consisting of a she (as the left hand electrode) and a rod of zinc dipping into a solution of Zn^{2+} ions at 298 K gave a measured emf of −0.789 V. What is the activity of the zinc ions?

SAQ 1.2e

Summary

This section deals with how cell potentials can be calculated as the difference between two standard electrode potentials. By convention, we always write the reduction reaction as the right hand half cell and the oxidation reaction as the left hand half cell.

Under non-standard conditions we must use the Nernst equation, which allows for activities other than unity. It is then possible to relate measured cell potentials to the activity of a given species in solution and in this way potentiometry can be used for quantitative analysis.

Objectives

You should now be able to:

- write chemical equations to represent the oxidation and reduction processes occurring at electrodes;
- describe galvanic cells as two half cells;

- sketch galvanic cells used for measuring standard electrode potential;
- use sep to predict the potentials of galvanic cells at unit activity;
- state the Nernst equation and use in calculation of cell potentials and solution activity.

1.3. LIQUID JUNCTION POTENTIALS

Overview

In this section we will see how two solutions of different concentrations, in contact can produce an emf. This emf will be measured along with E(cell) and will affect the accuracy of the analysis. We will also examine methods of minimising this possible source of error.

In the galvanic cells described in the earlier sections of this unit, the two half cells were linked together by a KCl bridge. An alternative method would be for the two solutions to be in contact, but separated by a porous division, such as sintered glass. This gives rise to two types of system with two and one liquid junction respectively. A liquid junction, as the name suggests, is the point of contact of two liquids across a membrane which allows electrical contact but prevents appreciable mixing. However, it is impossible to prevent mixing completely and a limited migration of ions across the boundary occurs. The consequence of this is the liquid junction potential.

Let us consider the liquid junction in the galvanic cell described by

$$\text{Pt} \mid \text{HCl}(a = 0.1 \text{ mol dm}^{-3}) \mathrel{\vdots} \text{HCl}(a = 0.01 \text{ mol dm}^{-3}) \mid \text{Pt}$$

which has only a single liquid junction. The cell is illustrated in Fig. 1.3a.

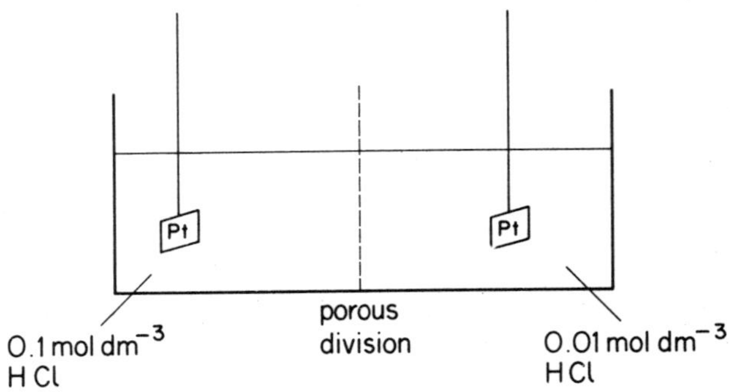

Fig. 1.3a. *Representation of a cell with a single liquid junction at a porous division*

Both the hydrogen and chloride ions will migrate across the boundary in both directions, but the net migration will be from the more concentrated to the more dilute. Hydrogen ions have a much higher mobility than chloride ions when placed in an electric field. Therefore, the migration of the two ions across the boundary will not be equal and, as the hydrogen ions are much faster, a greater proportion will pass across. This will result in a positive charge building up on the more dilute side of the boundary, leaving a negative charge on the more concentrated side. This is a separation of charge and a potential difference will quickly be created across the boundary. This is the liquid junction potential and must be allowed for in calculating the emf of a cell. The cell emf calculation can be rewritten as:

$$E(\text{cell}) = [E(\text{right}) - E(\text{left})] + E_j$$

where E_j is the liquid junction potential and can be a positive or negative quantity. In the above example $E_j = +40$ mV, which is a significant quantit'

If the liquid junction potential between 0.1 mol dm^{-3} and 0.01 mol dm^{-3} KCl is considered, the value of E_j is -1.0 mV. The reason for this lower value is that K^+ ions and Cl^- ions migrate at approximately the same rate. Thus, positive and negative ions arrive at the less concentrated side at about the same rate and cancel each other out.

If there are different species either side of the boundary, the tendency is for the more concentrated species to dominate. Consider the boundary between 0.05 mol dm^{-3} H_2SO_4 and 3.5 mol dm^{-3} KCl. The liquid junction potential is only $+4.0$ mV, even though the hydrogen ion migrates at a much faster rate than the sulphate. Thus, liquid junction potentials may be minimised by keeping a high concentration of KCl (or similar electrolyte with ions that migrate at equal rates) on one side of the boundary. As the concentration of the species on the opposite side of the boundary to the KCl increases then the liquid junction potential will also increase. This increase is particularly significant if the species in question contains hydrogen ions. The liquid junction potential is, therefore, pH dependent.

The significance of using a KCl bridge is to produce two opposite liquid junction potentials. These should largely cancel out, which should then minimise the total value of E_j. Saturated solutions of KCl are employed but if the presence of K^+ and Cl^- ions interfere with the analysis, suitable alternatives are ammonium chloride and potassium nitrate.

SAQ 1.3a

> Why is a liquid junction potential created between two solutions of hydrochloric acid of different concentrations?

SAQ 1.3a

SAQ 1.3b

Why are potassium chloride bridges used to minimise liquid junction potentials?

Summary

The concept of liquid junction potential between two liquids is described and its contribution to the total cell emf discussed. This can be minimized by interposing a saturated KCl solution.

Objectives

You should now be able to:

- describe the source of liquid junction potentials;
- describe how a KCl bridge can be used to minimise liquid junction potentials.

2. General Principles of Potentiometry

If you answered the questions in Part 1 correctly, you have the background knowledge and are ready to continue with this study book on potentiometry. If you were unsure of some of the answers, you may need to read the introductory study book on electrochemistry, *ACOL: Principles of Electroanalytical Methods*, before carrying on. You will then be in a better position to understand how potentiometry can be used to solve a large number of analytical problems.

In this part of the unit we will see how the basic principles of electrochemistry can be applied to potentiometry. We will then see how we can use this knowledge to obtain the concentration and/or activity of a species in solution from emf measurements.

Overview

In this section we will see how the Nernst equation can be applied to a galvanic cell used for measuring the activity of a given species in solution.

2.1. INTRODUCTION

Potentiometry is an extremely versatile analytical method and, as the name suggests, involves measuring the potential of a galvanic cell. Such cells consist of two half-cells and we can use the Nernst equation to calculate the potential of each. For example, consider the simple galvanic cell illustrated in Fig. 2.1a. This can be represented by

$$Zn(s) \mid Zn^{2+}(aq) \parallel Cu^{2+}(aq) \mid Cu(s)$$

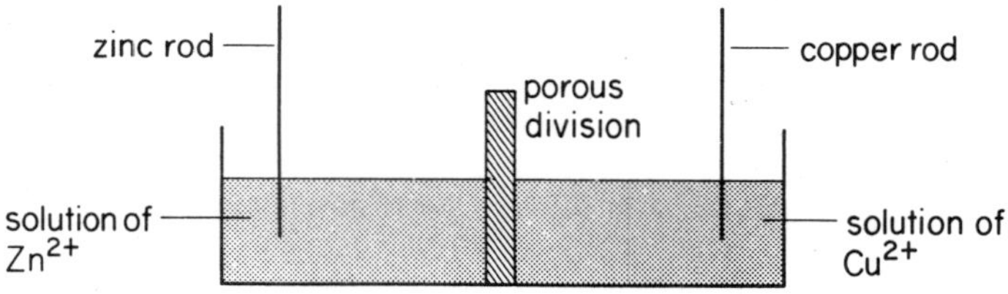

Fig. 2.1a. *Simple galvanic cell comprising Cu^{2+}/Cu and Zn^{2+}/Zn half cells*

Π Write the Nernst equation to calculate the potential of the right-hand half-cell.

The Nernst equation is:

$$E = E^{\ominus} - \frac{RT}{nF} \ln \frac{a(\text{reduced})}{a(\text{oxidised})} \tag{2.1a}$$

and for the copper/copper(II) half cell it is written as

$$E(Cu^{2+}/Cu) = E^{\ominus}(Cu^{2+}/Cu) - \frac{RT}{2F} \ln \frac{a(Cu)}{a(Cu^{2+})}$$

But we can take the activity of a pure substance to be unity, so

$$E(\mathrm{Cu^{2+}/Cu}) = E^{\ominus}(\mathrm{Cu^{2+}/Cu}) - \frac{RT}{2F} \ln \frac{1}{a(\mathrm{Cu^{2+}})}$$

or

$$E(\mathrm{Cu^{2+}/Cu}) = E^{\ominus}(\mathrm{Cu^{2+}/Cu}) + \frac{RT}{2F} \ln a(\mathrm{Cu^{2+}})$$

A similar expression could be written for the left-hand half-cell:

$$E(\mathrm{Zn^{2+}/Zn}) = E^{\ominus}(\mathrm{Zn^{2+}/Zn}) + \frac{RT}{2F} \ln a(\mathrm{Zn^{2+}})$$

The potential of the cell is then given by:

$$E(\mathrm{cell}) = E(\mathrm{Cu^{2+}/Cu}) - E(\mathrm{Zn^{2+}/Zn})$$

Clearly the potential of the cell will depend upon the activity of both the copper(II) and zinc(II) ions.

In chemical analysis, we generally wish to measure the activity or concentration of a single substance, rather than a combined value for two or more compounds.

Π In the above cell, what must we do to make the potential of the cell dependent on the activity of the copper(II) ions only?

I hope you thought of keeping the activity of the zinc(II) ions at a fixed value. In this way the potential of the left-hand half-cell is constant and:

$$E(\mathrm{cell}) = E^{\ominus}(\mathrm{Cu^{2+}/Cu}) + \frac{RT}{2F} \ln a(\mathrm{Cu^{2+}}) - E(\mathrm{Zn^{2+}/Zn})$$

As $E^{\ominus}(\mathrm{Cu^{2+}/Cu})$ and $E(\mathrm{Zn^{2+}/Zn})$ are both constant, they can be combined into one value, E'. Then:

$$E(\text{cell}) = E' + \frac{RT}{2F} \ln a(\text{Cu}^{2+})$$

and the measurement of cell potential is clearly proportional to the natural logarithm of the activity of the copper(II) ion.

This is typical of the majority of potentiometric measurements. The potential of one half-cell is kept constant and this entire half-cell is referred to as a reference electrode. Often the salt bridge that normally connects the two half-cells is also incorporated into the design. Reference electrodes are discussed in more detail in a later section and all that concerns us here is that they have a constant potential.

This now leaves us with the second half cell which contains the solution under investigation. Into this solution dips a suitable electrode.

Π What factors are important in selecting a suitable electrode for the analysis of a given species?

- It should have a Nernstian response to the activity of that species;
- It should not respond in any way to the activity of any other species present, ie it should be specific;
- It should not react chemically with any species present, ie it should be inert;
- The surface of the electrode should remain unchanged, even when small currents pass through the cell.

Few electrodes fit the requirements exactly, though a large number get very close and give more than acceptable results. The usual problem is that of specificity. Many electrodes are highly responsive to one substance, but are also responsive to other substances to a lesser extent. This behaviour is often acceptable and such electrodes are better described as selective, rather than specific.

Electrodes that respond to a specific ion are referred to as indicator electrodes and the selection and use of such electrodes is the key to potentiometry. Whilst a metal rod responds to its own ions in solution, it is not very selective as it also responds to a number of other metal ions. Similarly, a platinum electrode will respond to all redox couples in solution and again is not selective. The bulk of potentiometric measurements now involve specially designed electrodes which are collectively referred to as ion selective electrodes or ise for short. The most common of these is the glass electrode, which is selective to H^+ ions and, consequently, pH. The range of electrodes on offer covers a wide spectrum of species. Whilst the term ion-selective electrode is retained, many of these now respond to neutral species. Ion selective electrodes are available for the estimation of fluoride ions (F^-) in drinking water, glucose in blood, amino acids in biological fluids and a host of other determinations.

In our discussion on the zinc/copper galvanic cell, we have used the Nernst equation to relate the activity of the oxidised form of the metal to emf. However, it is possible to construct ion selective electrodes for both positive and negative ions, ie for both oxidised and reduced forms. It is, therefore, convenient to express the Nernst equation in a general way, which is applicable to all types of ise

$$E(\text{cell}) = E' \pm (RT/nF) \ln a_{\text{i}} \qquad (2.1\text{b})$$

where n is the charge on the ion, i, and E' is a constant incorporating the potential of the reference electrode and the standard potential of the half cell containing the solution under investigation and the ise. The other terms have their normal significance.

Note that the $\pm$ sign in the equation is used to signify that it will be positive if i is a cation and negative if i is an anion.

This equation would then apply to a typical cell used in potentiometric analysis and illustrated in Fig. 2.1b.

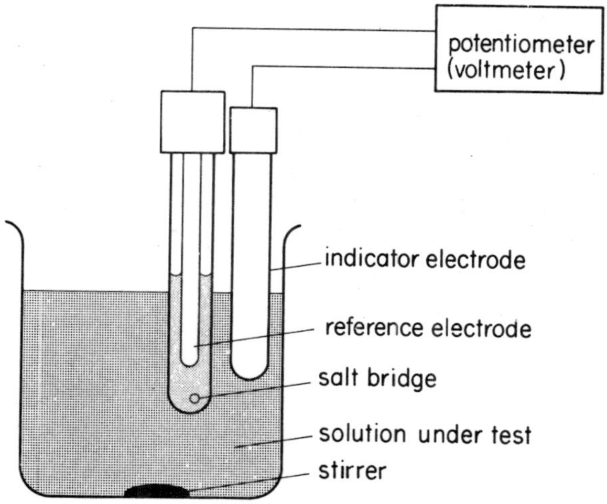

Fig. 2.1b. *Diagrammatic representation of a cell used for potentiometric analysis*

SAQ 2.1a

Which of the following describes the Nernstian response of a chloride ion selective electrode at 298 K?

(*i*) $E(\text{cell}) = E' + 0.0257 \ln a(\text{Cl}^-)$

(*ii*) $E(\text{cell}) = E' - 0.0257 \ln a(\text{Cl}^-)$

(*iii*) $E(\text{cell}) = E' + 0.0128 \ln a(\text{Cl}^-)$

(*iv*) $E(\text{cell}) = E' - 0.0128 \ln a(\text{Cl}^-)$

(*v*) $E(\text{cell}) = E' \pm 0.0257 \ln a(\text{Cl}^-)$

SAQ 2.1a

SAQ 2.1b

Draw a cell for the estimation of copper(II) ions by potentiometry.

Summary

In this section you have been introduced to indicator and reference electrodes, and to the use of galvanic cells for the determination of species in solution. We were then able to consider the idea of ion selective electrodes.

Objectives

You should be able to:

- write the Nernst equation in a form suitable for ion selective electrodes;
- describe an electrochemical cell suitable for potentiometric analysis.

2.2. IONIC STRENGTH ADJUSTORS

Overview

In this section, we will see how ionic strength adjustors can be used to allow the Nernst equation to be written with concentration, rather than activity values.

We have seen that the Nernst equation when applied to an ion-selective electrode in a potentiometric cell is written as:

$$E(\text{cell}) = E' \pm (RT/nF)\ ln\ a_i \qquad (2.1b)$$

As we generally quote activity and concentration measurements in multiples of 10, it is often more convenient to express the equation using logarithms to the base 10. As $\ln a = 2.303 \log a$ the Nernst equation then becomes:

$$E(\text{cell}) = E' \pm (2.303RT/nF) \log a_i$$

At 298 K this becomes:

$$E(\text{cell}) = E' \pm (0.0591/n) \log a_i \qquad (2.2a)$$

The proportionality constant between measured potential and logarithm of the activity of the ion is $(0.0591/n)$. In practice values of the proportionality constant between $(0.0550/n)$ and $(0.0591/n)$ are referred to as Nernstian. Strictly speaking, this is incorrect but electrodes with responses in this range are acceptable for analytical purposes, if properly calibrated. Values below $(0.0550/n)$ are termed sub-Nernstian and generally indicate a badly prepared electrode. However, this is not always the case as a limited number of electrodes do exhibit non-Nernstian behaviour. This proportionality constant is often referred to as the slope of the electrode.

Up to now, we have used the term activity of a solution, rather than the concentration and it must be stressed that ion-selective electrodes respond to the activity of the free ion in solution. This is in contrast to the majority of analytical methods which measure the total concentration (whether complexed or not) of a given species. Ion selective electrodes provide physical chemists with one of the few ways of obtaining direct activity measurements of specific ionic species in solution. Activity measurements are also of considerable use to biologists and biochemists, as a number of natural processes are dependent on activity, rather than on concentration.

However, analytical chemists generally require concentration values. The reason for this is partly historical and partly practical. In analytical chemistry it is often necessary to correlate the results of one technique with those of another and the early techniques were based on concentration measurements. Consequently, later techniques were also expected to produce 'understandable' concentration values. The second reason is that the preparation of standard solutions of known activity is difficult due to the problem of establishing the effect of all other species on the ion under investigation. The preparation of accurate concentration standards, on the other hand, is relatively straightforward. We, therefore, need to examine how an ise responds to changes in concentration as well as changes in activity.

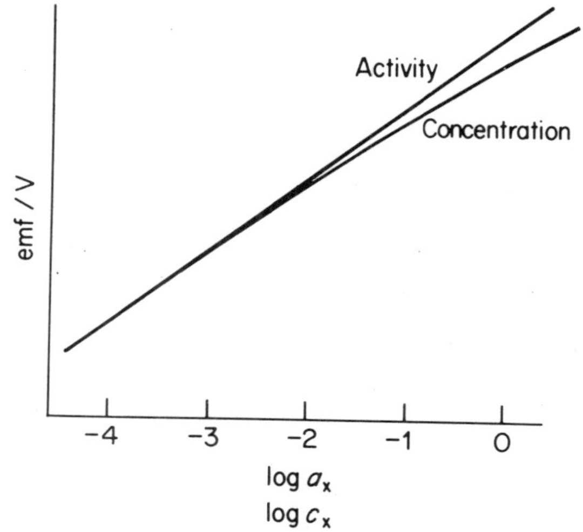

Fig. 2.2a. *Calibration curves for ion selective electrode using logarithm of activity and concentration of the anion of a pure 1 : 1 electrolyte*

Fig. 2.2a shows the response of an ion selective electrode to both activity and concentration standards of a pure binary electrolyte. If log(activity) is plotted against emf a straight line is obtained across the entire concentration range. This is in accord with the prediction of the Nernst equation. If log(concentration) is plotted against emf the graph is almost linear for low concentrations but clearly curves at higher concentration. For singly charged 1 : 1 electrolytes, such as NaCl, the two lines diverge above about 10^{-4} mol dm^{-3}, whilst for electrolytes containing ions with greater charge, such as $CuSO_4$, the two lines diverge at a lower value. The reason for this is that for very dilute solutions, the inter-ionic effects are minimal (ions well separated) and the activity approximates to concentration. As the ionic strength increases, inter-ionic attraction increases and deviations occur. If the solution contains ions other than those provided by the binary electrolyte, the deviations from linearity become even more significant for the log(concentration) graph.

Π Which of the two lines in Fig. 2.2a is the easiest to plot and why?

It is more convenient to use the straight line obtained from activity values. If we know the graph is a straight line a minimum of two calibration measurements are needed to plot it. In practice, it is better to use more points as this will increase the certainty that we are plotting the correct calibration line. Further, for straight line plots we can use statistical methods to obtain the best line. This also increases the probability of obtaining the correct answer in the analysis.

On the other hand, the curve obtained from concentration values requires a large number of points to give the shape and this would be time consuming for even the most routine of analyses.

Whilst the activity values give the straight line, the preparation of accurate activity standards is virtually impossible. To overcome these problems and obtain a straight line directly from concentration measurements, the procedure is modified. To each concentration standard and sample, a fixed ratio of an ionic strength adjustor or isa is added. This is a relatively concentrated solution of an inert electrolyte, ie one which does not react with the reagents in solution and does not participate in the electrode reaction. The theory behind its operation is discussed below.

The activity is related to the concentration by:

$$a_i = c_i \gamma_i \tag{1.1a}$$

The activity coefficient is a function of the total ionic strength (I) of all the ions present in the solution. For dilute solutions we can use the Debye Huckel relationship:

$$\log \gamma_i = -A z_i^2 I^{\frac{1}{2}} \tag{2.2b}$$

where A is a constant and z_i is the magnitude of the charge on the ion (ie generally 1 or 2). If the total ionic strength of all standards and samples were made the same, then γ_i would be a constant. In such cases Eq. 2.2a becomes:

$$E(\text{cell}) = E' \pm (0.0591/n) \log c_i \gamma_i$$

splitting the log term gives:

$$E(\text{cell}) = E' \pm (0.0591/n) \log \gamma_i \pm (0.0591/n) \log c_i$$

For a given ion, n is a constant and as γ_i is a constant for a fixed total ionic strength:

$$(0.0591/n) \log \gamma_i = \text{constant}$$

This constant can be combined with E' to give a new constant, E^*. The Nernst equation is then written as

$$E(\text{cell}) = E^* \pm (0.0591/n) \log c_i \qquad (2.2c)$$

Thus, at constant ionic strength the measured emf is directly proportional to log (concentration).

The ionic strength adjustor achieves this constant ionic strength as it is a relatively concentrated solution. When it is added to standards and sample the total ionic strength of the mixture is given by:

$I(\text{mixture}) = I(\text{isa}) + I(\text{sample})$

If $I(\text{isa}) \gg I(\text{sample})$

$I(\text{isa}) \approx I(\text{isa}) + I(\text{sample})$

$I(\text{mixture}) = I(\text{isa})$

Thus, all measurements are performed at the same total ionic strength and, consequently, constant activity coefficient. Fig. 2.2b shows the plot of emf against log(concentration) with and without an isa.

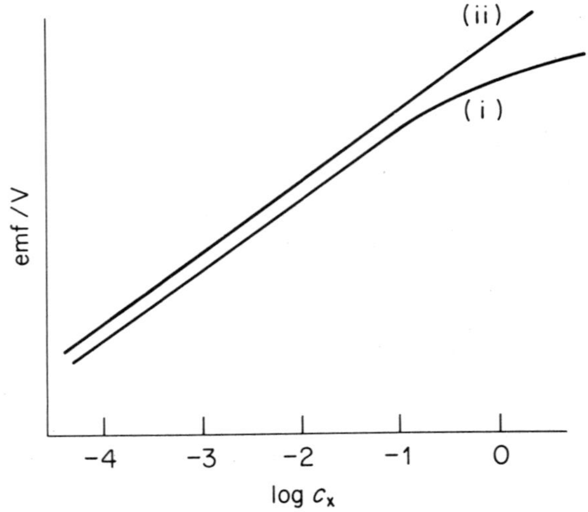

Fig. 2.2b. *Comparison of the emf against log (concentration) calibration plots (i) without and (ii) with an ionic strength adjustor*

SAQ 2.2a

> List two reasons why concentration standards are used in preference to activity standards when calibrating ion-selective electrodes. What conditions must exist when using concentration standards?

SAQ 2.2b

Select the appropriate phrase which best completes the sentence 'The purpose of adding an ionic strength adjustor to each sample and standard before each measurement is taken is to ...'

(*i*) ensure that all solutions have the same total ionic strength;

(*ii*) adjust the ionic strength to 1 mol dm^{-3};

(*iii*) allow emf measurements to be related directly to the activity of the ion under examination;

(*iv*) increase the measured emf giving more accurate readings.

Summary

The use of an ionic strength adjustor allows the Nernst equation to be written with concentration rather than activity. This permits the use of galvanic cells to determine concentrations of species in solution.

Objectives

You should now be able to:

- describe how ionic strength adjustors can be used to allow linear concentration plots for ion selective electrodes using concentration standards;
- list reasons why concentration standards are generally preferred to activity standards.

2.3. CALIBRATION

Overview

In this section we will see how ion selective electrodes can be calibrated using either activity or concentration standards.

Before you start working through this section on calibration, make sure you have a sharp pencil handy as you will need to plot a few graphs.

We have seen that we can use either activity or concentration standards to calibrate ion selective electrodes. As the two methods are very different, it is appropriate to consider them separately.

2.3.1. Activity Standards

We can easily prepare a concentration standard by accurate weighing of a primary standard. To convert this into an activity standard, we need a value for the activity coefficient and then use the relationship:

$$a_i = c_i \gamma_i \tag{1.1a}$$

Ion selective electrodes respond to the activity of a single ionic species and not to a pair of ions. This is one of the reasons why ion selective electrodes are useful in examining the behaviour of one type of ion in solution. However, whilst this feature is advantageous when we are performing the measurement, it is a major problem when we are carrying out the calibration. To prepare the activity standard for the individual ion we need to know its activity coefficient and not the mean activity coefficient of a pair of ions, ie we need to know γ_+ or γ_- and not $\gamma_\pm$. Other experimental electrochemical methods give a value for the mean activity coefficient and to obtain individual values from this require a few assumptions to be made. These assumptions, in turn, lead to some uncertainty in the accuracy of the activity standard. The degree of uncertainty depends upon the concentration range we are working in.

Π Using Fig. 2.2a as your reference, is there any range of concentration where determining γ_+ or γ_- is not necessary?

The answer to this question is fortunately yes. For 1 : 1 univalent electrolytes, such as NaCl, below a concentration of 10^{-4} mol dm^{-3}, the activity and concentration curves are superimposed. For divalent ions, the concentration will need to be somewhat lower for the concentration and activity values to be the same. Therefore, at very low concentrations we can assume that activity is equal to concentration, ie we assume that γ_+ or γ_- is unity. As most ion selective electrodes work to lower detection limits of at least 10^{-7} mol dm^{-3} – some work even lower – then there is quite a range where activity calibration is relatively easy. Having said that, one word of caution.

The activity coefficient is a function of the total ionic strength and will be affected by even seemingly inert ions. The above argument only applies to pure solutions and, consequently, activity standards must not contain any impurities.

It is with more concentrated solutions that our problems start. There are a number of different theories that can be used to estimate activities. However, it should be noted that all contain some approximations.

For solutions in the range 10^{-4} to 10^{-1} mol dm^{-3}, the Debye–Huckel limiting law provides a useful estimation of the activity coefficient. At lower concentrations the activity coefficients of both cations and anions can be calculated using the relationship

$$\log \gamma_i = -A z_i^2 I^{\frac{1}{2}} \qquad (2.2b)$$

whilst for more concentrated solutions we need to take account of the ionic radius and use an extended form:

$$\log \gamma_i = -A z_i^2 I^{\frac{1}{2}}/(1 + BdI^{\frac{1}{2}}) \qquad (2.3a)$$

Where B is a constant and d the effective diameter of the hydrated ion.

To a first approximation, we may calculate the activity coefficient of an individual ion in a solution which contains only the pure electrolyte.

When the ionic strengths of solutions are greater than 10^{-1} mol dm^{-3}, then the theories used become very much more complex. The activity coefficient is now not only dependent upon the total ionic strength, but also on the actual nature of the solution. Fortunately, we do not need to go into these theories in detail, merely use the results to help us calibrate our electrodes.

It is easier to summarise the consequences of the above treatments into a table. Fig. 2.3a gives the activity coefficients at different concentrations for three ions whose activities are often required:

Concentration /mol dm^{-3}	Activity coefficient K^+	Ca^{2+}	Cl^-
0.01	0.903	0.675	0.903
0.05	0.820	0.485	0.805
0.1	0.774	0.269	0.741
0.2	0.727	0.224	0.723
0.5	0.670	0.204	0.653
1.0	0.646	0.263	0.608

Fig. 2.3a. *Activity coefficients at different concentrations for K^+ (in KF), Ca^{2+} (in $CaCl_2$) and Cl^- (in KCL).*

Π When calibrating a Ca^{2+} ise, one of my students prepared three standard solutions of concentration 0.100, 0.500 and 1.000 mol dm^{-3}. What were the activities of the calcium ions in these three solutions?

Using the relationship (Eq. 1.1a)

$$a_i = c_i \gamma_i$$

The activities of the solutions are then

Concentration /mol dm^{-3}	Activity coefficient $\gamma(Ca^{2+})$	Activity /mol dm^{-3} $a(Ca^{2+})$
0.100	0.269	0.0269
0.500	0.204	0.102
1.000	0.263	0.263

The logarithm of these activities can then be plotted against the emf to obtain a calibration graph similar to that shown in Fig. 2.2a.

2.3.2. Concentration Standards

The measurement of concentration is by far the most common routine analytical procedure carried out using ion selective electrodes. Fortunately, a concentration standard may be prepared by weighing any suitable primary standard. Other standards are then prepared by standard dilution. Whilst there are several ways of obtaining concentration values using ion selective electrodes, we are only concerned here with the preparation of a calibration graph.

Π Ion selective electrodes give a straight line calibration plot when the logarithm of the activity of the ion is plotted against emf. How do we obtain a straight line calibration graph using concentration values in place of activity?

I hope you remembered that we add an ionic strength adjustor as described in Section 2.2. Ionic strength adjustors, or isa for short, keep the activity coefficient constant so that the Nernst equation at 298 K becomes:

$$E(\text{cell}) = E^* \pm (0.0591n) \log c_i \qquad (2.2c)$$

We may then plot a graph of measured emf against log (concentration).

However, looking up the logarithm concentration each time can be a bit tedious, so we use a quicker method.

We can buy a special type of graph paper called semi-log (or log/mm) paper. This has one non-linear axis as shown:

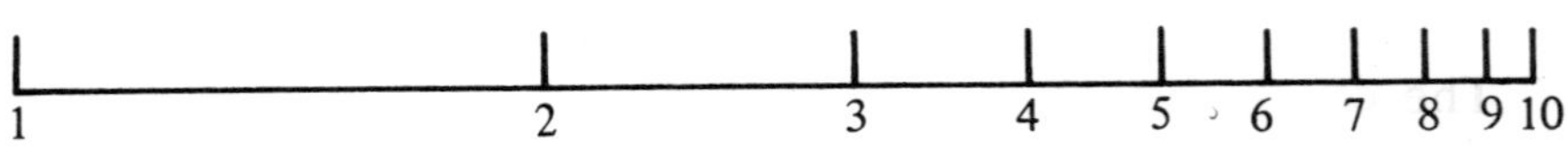

This axis is prepared by taking the log of the number to find its relative position on a scale between 0 and 1. For example, the log of 3 is 0.477 and examination of the scale shows that 3 is roughly half way between 1 and 10. Thus, we no longer need to look up the log value of this logarithm in tables, we simply plot it on a log axis.

Semi-log paper comes in one cycle, two cycle, three cycle etc. Each cycle is an exact repetition of the single cycle shown above and corresponds to an order of magnitude or a decade. For example, if we were using three standards 1.0×10^{-3}, 2.0×10^{-3} and 5.0×10^{-3} mol dm^{-3}, then we would use one cycle graph paper as all three are the same order of magnitude. If the three standards spanned a wider concentration range then we would need to use more cycles.

Π If the standards used to calibrate an ise had concentrations of 1.0×10^{-3}, 5.0×10^{-3} and 2.0×10^{-4} mol dm^{-3}, which type of semi-log graph paper (1, 2 or 3 cycles) would be required?

You would need two-cycle paper as your standards span two orders of magnitude. The values 1.0×10^{-3} and 5.0×10^{-3} would be plotted on one cycle and 2.0×10^{-4} on the next.

Let's try an example together. A chemist in a local water control laboratory calibrated a fluoride ise, with the following results

Concentration of F^- /mol dm^{-3}	emf /mV
5.00×10^{-5}	−122
1.00×10^{-4}	−106
5.00×10^{-4}	− 66
1.00×10^{-3}	− 49
5.00×10^{-3}	− 10

The concentrations span three orders of magnitude, so we chose three cycle semi-log paper. The calibration graph is then shown in Fig 2.3b. The three cycles of the log axis are labelled as 10^{-5}, 10^{-4} and 10^{-3} so that you can see each order of magnitude clearly. The emf is plotted on the linear axis in the normal way. We are fortunate in this case that the line goes through all the points. In practice, some points are just off the line and we have to draw the 'best straight line' or use a mathematical model such as 'least squares' to help us.

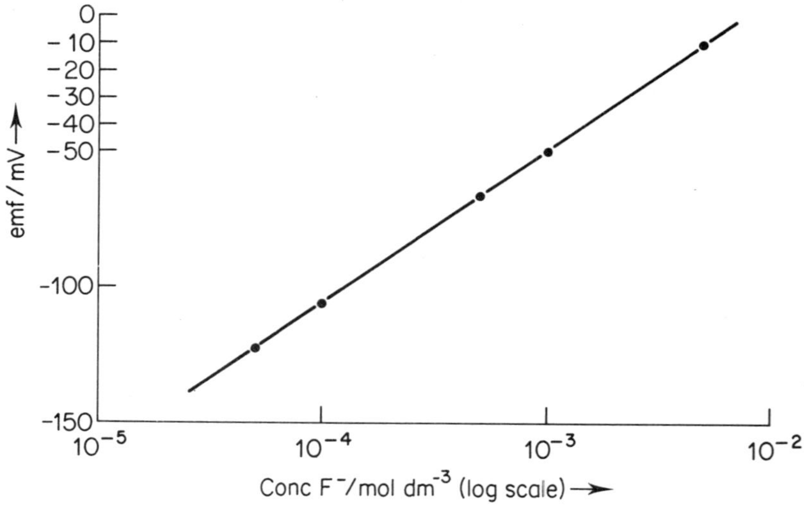

Fig. 2.3b. *Calibration curve for a fluoride ion-selective electrode*

SAQ 2.3a

A student was calibrating a calcium ise and obtained the following results. What is the concentration of a sample (marked S) which was measured at the same time?

Concentration of Ca^{2+} /mol dm^{-3}	emf /mV
1.00×10^{-4}	-2
5.00×10^{-4}	$+16$
1.00×10^{-3}	$+25$
5.00×10^{-3}	$+43$
1.00×10^{-2}	$+51$
S	$+33$

SAQ 2.3a

SAQ 2.3b

Which of the following is *not* a reason why concentration standards are generally used in potentiometry?

(*i*) Concentration standards may be prepared from a primary standard.

(*ii*) The calibration graph of log (concentration) against emf is always a straight line.

(*iii*) Concentration values facilitate comparison with other techniques.

(*iv*) Accurate activity standards are difficult to prepare as values of the activity coefficient are the result of assumptions.

SAQ 2.3c

Use Fig. 2.3a to estimate the activity of the potassium ion in a 0.200 mol dm^{-3} solution of potassium chloride.

Summary

The construction of calibration plots of emf against log (concentration) have been described and a method of plotting in semi-log graph paper has been explained.

Objectives

You should now be able to:

- calculate the activity of standard solutions from a table of activity coefficients;
- draw a calibration graph for concentration standards.

2.4. SELECTIVITY

Overview

Up to now we have considered only the influence of a specified ion on an ise. In this section, we will see how the Nernst equation can be modified to allow for interfering ions and calculate the maximum level of interference we can tolerate for a given accuracy in our analysis.

Whilst an ise should have a Nernstian response to a specific ion, it is likely to respond to a greater or lesser extent to other ions present in the solution. This is particularly true if the other ion is physically and chemically similar to it. For example, the chloride (Cl^-) ise suffers considerable interference from the similar bromide (Br^-) ion, but suffers much less interference from the not so similar hydroxide (OH^-) ion.

How can we quantify the interference of an ion on an ise? We can express the degree on interference very simply. For example, the EIL chloride ise responds 300 times more to bromide ions than it does to chloride ions, ie the activity of the bromide ions has the same effect as 300 times the same activity of chloride ions. We could rewrite the Nernst equation (Eq. 2.1b) for the chloride and reference electrodes as:

$$E = E^* - (RT/F) \ln [a(Cl^-) + 300a(Br^-)]$$

(Remember the negative sign because we have a negative ion.)

On the other hand the same electrode responds to hydroxide ions 100 times less than it does to chloride ions. In this case the Nernst equation becomes:

$$E = E^* - (RT/F) \ln [a(Cl^-) + 0.01a(OH^-)]$$

All we have to do is to multiply the activity of the interfering ion by a factor which we call the selectivity coefficient or selectivity ratio. In the above examples the magnitude of the charge on the interfering ions is the same as on the ion under investigation, in this case 1. To extend the equation to the general case we need to write the log term in the Nernst equation in a different way. The theory of logarithms gives:

$$y \ln x = \ln x^y$$

so that

$$(1/n) \ln a = \ln a^{1/n}$$

for an ion y which interferes with an electrode selective for an ion i, we then write:

$$E = E^* \pm (RT/F) \ln [a_i^{1/n} + k_{iy}\, a_y^{1/q}] \qquad (2.4a)$$

where k_{iy} is the selectivity coefficient of the interfering ion y over the ion under investigation i, whilst q is the magnitude of the charge on the ion y. Clearly, the greater the magnitude of k, the greater the problem with interference from the second ion.

If there is more than one interfering ion, then we can sum the products of the selectivity coefficients and activities of each of the ions. The general equation then reaches its final form

$$E = E^* \pm (RT/F) \ln [a_i^{1/n} + \sum_j k_{ij} a_j^{1/q}] \qquad (2.4b)$$

For example, when using the chloride ise in the presence of both bromide and hydroxide ions, Eq. 2.4b becomes:

$$E = E^* - \frac{RT}{F} \ln [a(Cl^-) + 300a(Br^-) + 0.01a(OH^-)]$$

It should be noted, however, that the validity of expanding such an equation over more than two interfering ions is questionable.

Π Calculate the maximum activity we can tolerate for the interfering bromide ion, if the chloride ise is to be used to analyse within a 5% error, solutions whose chloride activities are of the order of 10^{-3} mol dm^{-3}.

$$k_{Br,Cl} = 3.00 \times 10^{2}.$$

$$\text{Error in measurement} = \frac{\text{effect of } Br^- \text{ on the electrode}}{\text{effect of } Cl^- \text{ on the electrode}} \times 100$$

$$= k \times \frac{a(Br^-)}{a(Cl^-)} \times 100$$

$$3.00 \times 10^{-2}\, a(Br^-) \times 100/10^{-3} = 5$$

Thus $a(Br^-) = 1.67 \times 10^{-3}$ mol dm^{-3}

This value for the activity of the bromide ion is the maximum we can tolerate to obtain a result within 5% of the expected value when the activity of the chloride ion is 10^{-3} mol dm^{-3}. If the activity of the chloride ion is higher than 10^{-3} mol dm^{-3} then we can tolerate a greater activity of bromide ions. In fact the important parameter is the ratio of the activities. Examination of the above calculation shows that:

$$\text{error in measurement} = 100 \times k \times a(Br^-)/a(Cl^-)$$

or

$$a(Cl^-)/a(Br^-) = 100k/\text{error} \tag{2.4c}$$

or if the activity coefficients are the same for both ions

$$c(Cl^-)/c(Br^-) = 100k/\text{error} \tag{2.4d}$$

We can, therefore, calculate the ratio of the concentrations of the analyte and interfering ions for a given error. If the activity of the interfering ion exceeds the value allowed by this ratio, we have four options:

(*i*) if the activity of the interfering ion is known accurately, then we can make allowances for it in the calculation;

(*ii*) if we are preparing a calibration graph, we can make the interfering ion activity of the standards and samples the same. This will give a constant level of interference and not affect the accuracy;

(*iii*) we can remove the interfering ion by precipitation or complexation. However, we would have to be very careful that we didn't remove the ion under investigation as well;

(*iv*) we can bring the ratio of ion under investigation to interfering ions within the allowed range by adding a known amount of the former. This effectively 'swamps' the interfering ion. This is the principle of the technique of multiple standard addition which is discussed later.

Option (*i*) is not a very accurate method as selectivity coefficients can vary with the total ionic strength of the solutions. Thus, seemingly non-interfering ions, such as those present in the ionic strength adjustor, will alter the value of k. The use of selectivity coefficients is, consequently, more for guidance in calculating maximum levels of interference than in quantitative calculations.

SAQ 2.4a

Calculate the maximum tolerated activity of silver ions for a 10% error when using a potassium ise to measure K^+ ions in the range 10^{-3} to 10^{-4} mol dm^{-3}. $k_{K,Ag} = 1.00 \times 10^{-4}$.

SAQ 2.4b

Examine each of the following statements and decide whether it is true or false.

(*i*) Ion selective electrodes respond to ions other than the ion under investigation.

(*ii*) A large value for the selectivity coefficient indicates a highly interfering ion.

(*iii*) Selectivity coefficients are constant across the entire concentration range.

(*iv*) Selectivity coefficient is the ratio of an electrode's response to an interfering ion relative to the ion under investigation.

Summary

The Nernst equation has been identified for non-selective electrodes to allow for the presence of interferring ion. The maximum interference which can be tolerated for a given accuracy has been determined.

Objectives

You should now be able to:

- state that ise's respond to more than one ion;
- define the term selectivity coefficient;
- use selectivity coefficients to modify the Nernst equation to compensate for interfering ions;
- calculate the maximum allowed level of interference.

2.5. LIMITS OF MEASUREMENT AND RESPONSE TIMES

Overview

In selecting an ise for a given task, we need to know as much about the electrode as possible. Not only do we need to know about possible interferences, but also the concentration range over which the electrode works and the time it takes to respond.

On the surface, it would appear easy to quote a single limit of measurement for an electrode and the time taken for it to respond. However, there are a number of different definitions of both limit of measurement and response time.

How do we define the lower limit of measurement? The lower limit of measurement is generally the quantity that restricts the use of an ise. If a solution is more concentrated than the upper limit, it is a relatively easy process to dilute it and bring it into the normal working range.

The lower limit of measurement is not the same as the lowest detection limit. Whilst an ise may produce an emf output in almost any solution, we need to be able to interpret the signal. Fig. 2.5a shows the response of a typical ise over a wider concentration range than previously considered.

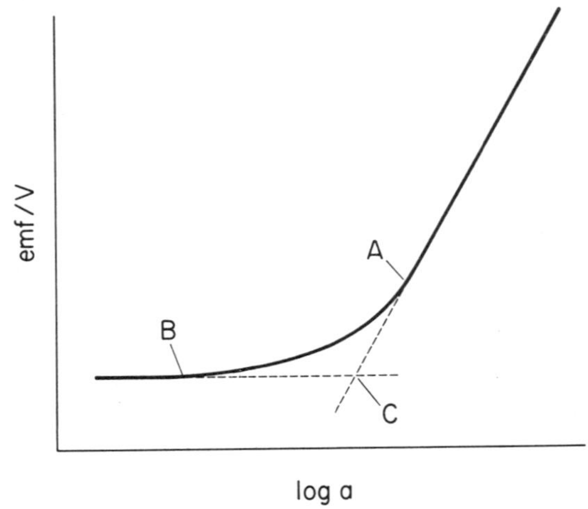

Fig. 2.5a. *Calibration graph for a typical ion-selective electrode*

We can see that the straight line calibration becomes curved at point A. This is the lower limit of Nernstian response and many quote this point as the lower limit for useful measurement.

Whilst the calibration graph is curved below A, it can still be used for measurement. It is only when the graph becomes almost horizontal at point B that measurement of ionic concentration become impossible. This is recognised in the IUPAC recommended definition of lower detection limit which states:

> *The lower detection limit is the concentration of the ion under investigation at which the extrapolated linear portion of the calibration graph at extreme dilution of that ion intersects the extrapolated Nernstian portion of the graph.*

This is indicated by point C on Fig. 2.5a. The plotting of an accurate curved calibration line below the limit of Nernstian response requires a large number of points and will be time consuming. Further, certain methods for performing analyses using ises use the value of the electrode slope in the calculation. Such methods cannot be used with any degree of certainty in the curved region.

To summarise, if you don't mind the extra work in calibrating the electrode, the IUPAC definition tells you how low you can measure, whilst the limit of Nernstian response tells you lowest concentration you can measure easily.

In the first instance, both of these lower limits will be determined by the construction of the electrode itself. For example, the chloride (Cl^-) ise uses a solid state membrane of silver chloride, which will be in contact with the solution. Silver chloride is limitedly soluble

$$AgCl(s) \rightleftharpoons Ag^+(aq) + Cl^-(aq)$$

The solubility product of this is such that at 25 °C the concentration of chloride ions will be in the region of 10^{-5} mol dm^{-3}. Thus, unless an excess of silver ions are present to force the equilibrium to the left, the electrode cannot be used to measure concentrations of Cl^- below 10^{-5} mol dm^{-3} or the membrane will dissolve to bring the concentration up to the required level.

Π How can we reduce the lower limit?

There are two common ways. Firstly, the above solubility refers to aqueous solution. If we change the solvent, then we will have a different value for the solubility product. Secondly, we can make use of the le Chatelier principle. The enthalpy of solution of silver chloride is positive. Thus, it will be less soluble in cold water. For example at 5 °C, the chloride ion concentration drops to around 10^{-6} mol dm^{-3}.

Π What other factors influence the limits of measurement?

There are a number of factors to consider. I hope you thought of an interfering ion which cannot be eliminated. The limit of accurate measurement will be in a ratio to the level of interference and will be given by Eq. 2.4c, 2.4d.

Other factors include the purity of any standards that we need to use to calibrate the system and also the purity of the solvent itself. For example, the limit of measurement of the gas sensing ammonia probe is governed by the purity of the water. It is very difficult to obtain water with an ammonia concentration less than about 10^{-7} mol dm^{-3}. All standards will contain this much ammonia and it effectively becomes our lower limit.

There are, clearly, a number of factors that influence the lower detection limit and it is virtually impossible to predict a value from theoretical considerations. The only effective way to ascertain the limit is to plot the calibration curve under the conditions that will prevail during the analysis.

How do we define response time? Ion selective electrodes generally respond very quickly to changes in concentration and steady readings are obtained within seconds of the electrode starting to respond to a change in concentration. The IUPAC definition of response time is unambiguous.

The response time is the time taken for the potential of the cell to reach a value 1 mV from the final equilibrium potential.

This can be seen in Fig. 2.5b, which shows the variation in the response time of the Orion ammonia gas sensing probe with concentration. This is a relatively sluggish electrode in terms of response time, but for this reason, it is easier to monitor.

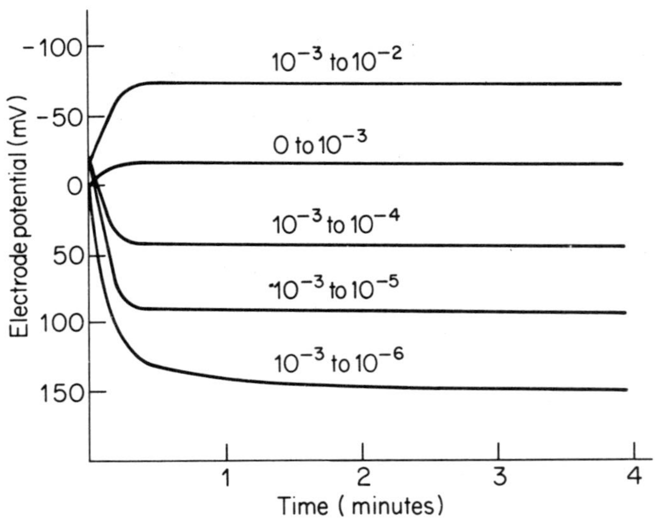

Fig. 2.5b. *Typical electrode response to step changes in ammonia concentration (concentration ranges are mol dm^{-3} NH_3)*

The factors influencing response times include:

– the type of membrane; glass and solid state membranes tend to respond more quickly than do liquid ion exchange membranes or gas sensing probes. Having said this, the slowest membranes have generally responded within about 30 s of being subjected to a change in concentration;

– the magnitude of the concentration change. Fig. 2.5b shows that it takes marginally longer for the electrode to respond to a large concentration change than to a small one;

– the total volume of the test solution and the rate of stirring. Generally small volumes and relatively fast stirring rates will decrease the response time;

– how the change in concentration is brought about. If the electrode is transferred between two beakers, which contain the same ion, but at different concentrations, the electrode will respond rapidly. This is because the new solution is homogeneous and the part of the solution next to the electrode will be representative of it all. On the other hand, if the electrode is immersed in a solution which is gradually diluted by slow addition of the solvent, the response time will be much longer. In this case, the time taken to obtain a homogeneous solution must be added to the response time of the electrode itself;

– interfering ions generally increase the response time;

– temperature; as with the majority of chemical processes, the response time is likely to decrease with an increase in temperature;

– the above discussion has revolved around the electrode itself. The cell emf is displayed on a voltmeter which is often damped to give steadier readings. Whilst the electrode may have come to equilibrium, the meter may not indicate it immediately.

As with limits of measurement, response times are best determined empirically.

SAQ 2.5a

Fig. 2.5c is a calibration graph for an Orion nitrate (NO_3^-) ise.

Indicate on the graph:

(*i*) the limit of Nernstian response;

(*ii*) the lower limit of measurement as defined by IUPAC.

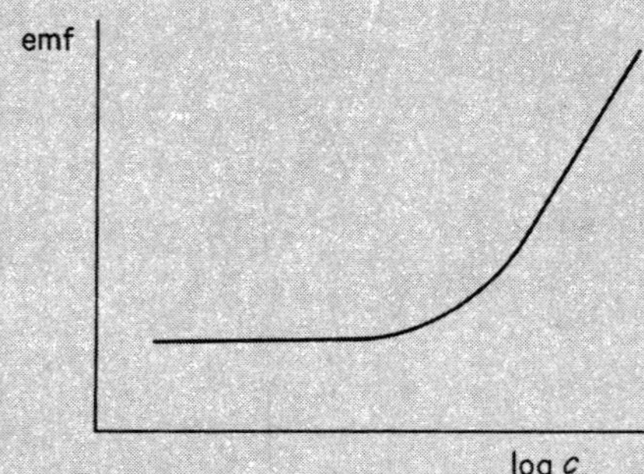

Fig. 2.5c. *Calibration graph for nitrate ise*

SAQ 2.5b

Which of the following factors influence the response time of an ise?

(*i*) temperature;

(*ii*) interfering ions;

(*iii*) the voltmeter;

(*iv*) type of electrode;

(*v*) stirring rate.

Summary

The limit of measurement, and the response time of an ion selective electrode are defined and the factors affecting these discussed.

Objectives

You should now be able to:

- define the limits of measurement and calculate them from calibration graphs;
- define response time and list the factors that influence it.

2.6. COMPARISON OF ANALYTICAL METHODS

Overview

In this section we will consider different methods for performing measurements using ion selective electrodes. We can group these into direct reading methods, incremental methods and potentiometric titrations. As a rough guide, as we move down the list the methods become more time consuming but can be more accurate.

2.6.1. Direct Reading Methods

The basis of this technique is the plotting of a calibration graph as described in 2.3. The samples are then treated in the same way as the standards and the concentration (or activity) read directly off the calibration graph. Once the calibration graph is set up, a competent technician can analyse about 30 samples an hour. If an automated process is used, then this figure can be increased to around 120.

Π What requirement do we have of response time if the electrode is to measure 120 samples in one hour?

120 samples per hour equates to one sample every 30 s. The contact time between electrode and solution may be even less than this because a wash cycle may be needed between measurements to prevent sample carry over.

Therefore, we are looking for a response time better than 15 s or so.

Under ideal conditions, where temperature, stirring, sample pretreatment etc are carefully controlled, accuracies of $\pm 1\%$ are possible using the direct reading method. Under normal laboratory conditions $\pm 2\%$ is a more realistic figure for monovalent ions.

Π Would you expect the accuracy of an ise for a monovalent ion to be the same as for a divalent ion?

Unfortunately, the accuracy decreases with the increasing charge on the ion. This is because the Nernstian slope for a monovalent ion is 59.1 mV per decade but for a divalent ion it is 29.6 mV per decade. The latter is a 'flatter' graph giving a greater uncertainty in measuring concentration for a given emf. Accuracies in measuring divalent ions are, therefore, of the order of $\pm 4\%$.

The direct reading method of analysis is quick, once the calibration graph is plotted. Recalibration is advisable, though the frequency at which it occurs depends on the electrode construction. Once a day is quoted for glass and solid state electrodes whilst once an hour is recommended for liquid ion exchange and gas sensing probes.

Π How can we simplify the calibration process?

This is something you have probably done yourself if you have calibrated a pH meter with a buffer solution. The entire system is calibrated with two standard solutions or, in some cases, even one. These are termed two point and one point calibration respectively. Let's examine the Nernst equation (again) under conditions of constant ionic strength for the glass-reference electrode assembly, which is selective for H^+ ions. When we write it we will represent the Nernstian slope of the calibration line by S, ie $S = 2.303RT/F$.

$$E(\text{cell}) = E^* + S \log c(\text{H}^+) \qquad (2.6a)$$

Note that the charge on the ion, *n*, is unity. Now:

$$\text{pH} = -\log c(\text{H}^+)$$

Eq. 2.6a then becomes:

$$E(\text{cell}) = E^* - S\ \text{pH}$$

When we measure *E*(cell) for a buffer solution of known pH, then if we assume Nernstian slope, the only unknown quantity in the above equation is E^*. We then calibrate the meter for this value. pH meters are in effect voltmeters calibrated in pH units.

Π What change in measured emf corresponds to a change in pH of one unit at 298 K?

Consider two measurements where the measured emf is E_1 at pH_1 and E_2 at pH_2.

Then:

$$E_1 = E^* - S\ \text{pH}_1$$

$$E_2 = E^* - S\ \text{pH}_2$$

Subtracting these equations gives:

$$E_1 - E_2 = S\ (\text{pH}_2 - \text{pH}_1)$$

For a change in pH of one unit

$$E_1 - E_2 = S$$

$$= 2.303RT/F$$

$$= 2.303 \times 8.314 \times 298/96487$$

$$= 0.0591\ \text{V}$$

therefore

$$E_1 - E_2 = 59.1 \text{ mV}$$

Once set at a given pH, for every decrease of 59.1 mV, the meter indicates an increase in pH of one and vice versa.

If the temperature is not 298 K, the value of S will change. Consequently, many pH meters either have a temperature probe which automatically compensates for changes in temperature or it must be done manually using a temperature control on the front of the meter. We can extend this calibration with one standard to any ise having Nernstian response. Direct reading ion meters are available with the voltage scale calibrated in concentration units. Such meters have a switch that allows selection of single or doubly charged ions – remember the slope for the doubly charged ion is half that of the singly charged.

Π We have identified that temperature might be a problem if we are using one point calibration, can you think of any others?

There are two that readily spring to mind.

Firstly the equation (Eq. 2.6a) as written assumes constant ionic strength. This should not be a problem if an ionic strength adjustor is present, but with certain pH measurements this is not the case. It is, therefore, essential that the meter is calibrated with a buffer solution in the range that you are working in, ie if you are measuring acid samples in the region pH 1 to 2, then do not calibrate with a buffer solution at pH 7 and assume the meter is calibrated across the complete range.

Secondly, the meter is programmed for values of S = 59.1 mV, ie for true Nernstian behaviour. In practice, electrodes can deviate from this due to a number of reasons such as contamination of the surface and interfering ions. To allow for this we make use of a two point calibration. If two points are known we can plot a calibration graph. Alternatively, some ion meters will perform their own internal calculation of slope from the two measurements and then output sample concentrations directly. Life does get easier with these modern meters but unfortunately they are a bit more expensive.

SAQ 2.6a

Which of the following statements are true?

(*i*) The direct reading method is quick once the electrode is calibrated.

(*ii*) The accuracy of analysis using ion-selective electrodes is the same for singly and doubly charged ions.

(*iii*) Calibration of ion selective electrodes requires a minimum of two standards.

(*iv*) A pH meter is in effect a voltmeter calibrated in pH units. A decrease in measured emf of 59.1 mV always corresponds to an increase in pH of one unit.

(*v*) pH meters should always be calibrated in the same pH range as that of the sample(s).

2.6.2. Incremental Methods

These methods all involve adding a reagent to the sample and measuring the change in cell potential. The added reagent can be a standard solution of the specific ion under study, in which case the technique is one of standard addition. Alternatively, the reagent can quantitatively react and remove a fixed amount of the ion, in which case, the technique is one of standard subtraction.

The most common incremental method is standard addition. Here we start with a solution of the ion and the cell potential will be given by:

$$E_1 = E^* \pm S \log c \qquad (2.6b)$$

where c is the concentration of the ion. If we now add to this a solid reagent which, when dissolved, increases the concentration of the ion by an amount x, the measured emf will be given by:

$$E_2 = E^* \pm S \log (c + x) \qquad (2.6c)$$

Subtracting Eq. 2.6b from 2.6c gives:

$$E_2 - E_1 = \pm S \log (c + x) \pm S \log c$$

$$= \pm S \log [(c + x)/c]$$

which on rearranging becomes:

$$c = x/\{\text{antilog}\, [(E_2 - E_1)/ \pm S] - 1\} \qquad (2.6d)$$

All the quantities on the right hand side of the equation can either be measured or are known theoretically. Hence, the concentration of the ion in the solution under investigation can be found. As the concentrations represented by c and x are ratioed, they need not necessarily be in mol dm^{-3}. Any suitable units, such as ppm, can be used. The important point being that c and x must be in the same units.

Π What happens to the above calculation if the standard addition is made by adding a known volume of a standard solution of the ion to the sample?

If the volume of addition is small, say less than 10% of the sample, then we can neglect the effect of diluting the sample. The equation obtained will then be identical to Eq. 2.6d and the quantity x will be the concentration of the standard after it has been added to the sample.

If the original sample volume is small, then the addition is likely to result in significant dilution of the sample. In this case we must allow for the change in concentrations. If the original volume of the sample is V_1 cm^3, that of the addition is V_2 cm^3 and the concentration of the standard solution prior to addition is x, then:

concentration of original sample after addition =

$$\frac{c\,V_1}{(V_1 + V_2)}$$

concentration of standard after addition =

$$\frac{x\,V_2}{(V_1 + V_2)}$$

The equations representing the two emf measurements are now

$$E_1 = E^* \pm S \log c$$

and

$$E_2 = E^* \pm S \log [(cV_1 + xV_2)/(V_1 + V_2)]$$

subtracting and rearranging as before gives:

$$c = xV_2/\{(V_1 + V_2) \text{ antilog } [(E_2 - E_1)/ \pm S] - V_1\} \quad (2.6e)$$

Let's try an example.

Π A copper (Cu^{2+}) ise/ref system gave a reading of 110 mV when immersed in 50 cm^3 of a solution of copper(II) ions at 298 K. After addition of 5.00 cm^3 of a 0.100 mol dm^{-3} standard solution of copper(II) the measured emf changed to 130 mV. Calculate the concentration of the original copper(II) solution.

For a copper(II) ise, $n = 2$ and the slope will be positive. ie

$$S = +RT/2F.$$

Substituting in Eq. 2.6e gives

$$c = \frac{0.100 \times 5.00}{[(50.0 + 5.00)\text{antilog}\,(130 - 110)10^{-3}/0.0296] - 50.0}$$

$$c = 2.4 \times 10^{-3} \text{ mol dm}^{-3}$$

Π What assumptions are made in the above calculation?

There are five significant ones:

- there is Nernstian response over the range of the addition;
- the total ionic strength remains constant;
- there is no change in the liquid junction potential between the reference electrode and the solution;
- there is no temperature change on mixing;
- the standard addition contains no impurities.

These assumptions are generally valid providing the magnitude of the addition is not too great. However, whilst we can measure x, V_1 and V_2 very accurately, the major source of error is in the measurement of emf or, more precisely, the difference in emf. We can measure differences in emf to ± 0.2 mV and to obtain relative errors less than 5%, we need differences in emf of at least 4 mV. In practice, differences in the order of 10 to 30 mV are preferred and the magnitude of the standard addition is a compromise between maintaining the above assumptions and obtaining sufficient difference in emf. Thus, it is generally better to add a small volume of a concentrated solution than a large volume of a dilute solution.

Another point worth noting at this stage is that the electrode slope (S) plays a part in the calculation. Whilst we can calculate it from theoretical considerations, it is better to use values obtained experimentally from calibration plots. Most electrode manufacturers supply slope values for their electrodes at different temperatures for use in this type of calculation. Alternatively you could obtain your own.

An extension of the standard addition technique is the multiple standard addition method. If we ignore the effect of dilution when we add V_S cm^3 of a standard solution of concentration x to V cm^3 of a sample, the emf will be given by:

$$E(\text{cell}) = E^* \pm S \log (c + V_S x/V)$$

Dividing by S and taking antilogs gives:

$$\text{antilog}\,[E(\text{cell})/S] = \text{constant}\,(c + V_S x/V)$$

where the constant is antilog (E^*/S). All the quantities in this equation except for E(cell) and V_S are constant. Hence:

$$\text{antilog}\,[E(\text{cell})/S] \propto V_S$$

and a plot of antilog $[E(\text{cell})/S]$ against V_S will yield a straight line as shown in Fig. 2.6a.

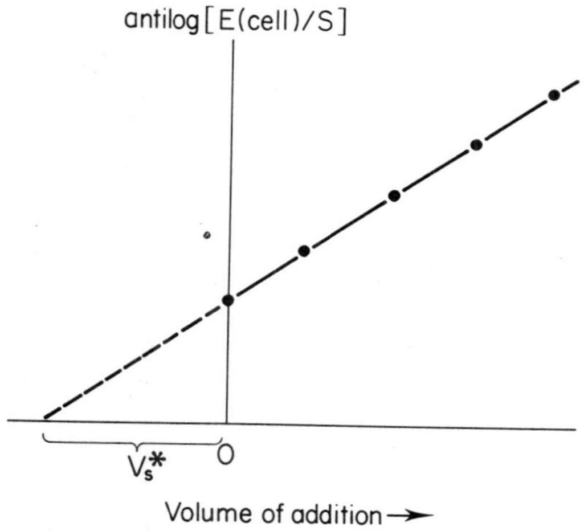

Fig. 2.6a. *Typical plot for multiple standard additive procedure (The straight line is extrapolated to the abcissa and a value for* V_s^* *read off)*

If we extrapolate this line back to the abcissa (the volume axis in this case) the intercept can be used to calculate the concentration of the sample.

When the antilog $[E(\text{cell})/S] = 0$, since the constant has a finite value, then:

$$c + V_s^* x/V = 0$$

thus:

$$c = -V_s^* x/V \qquad (2.6f)$$

The value of $-V_s^*$ is read off the axis and the concentration of the original solution calculated.

Π To 100 cm^3 of a sample of potassium ions of unknown concentration was added known volumes of a standard 0.100 mol dm^{-3} solution of K^+ ions. The emf of the ise/ref combination was measured at 298 K after each addition. The results are tabulated:

Volume K^+ solution	emf/mV V_s/cm^3
0	136
1.00	150
2.00	160
3.00	166
4.00	172
5.00	176

Calculate the concentration of the original solution.

Assuming a Nernstian slope of 59.1 mV for a monovalent cation we can calculate antilog(E/S)

E/mV	E/S	Antilog(E/S)	V_s/cm^3
136	2.300	199	0.00
150	2.537	344	1.00
160	2.706	508	2.00
166	2.807	641	3.00
172	2.909	810	4.00
176	2.976	947	5.00

The plot of antilog(E/S) against V is shown in Fig. 2.6b. The intercept on the abcissa occurs at -1.34 cm^3.

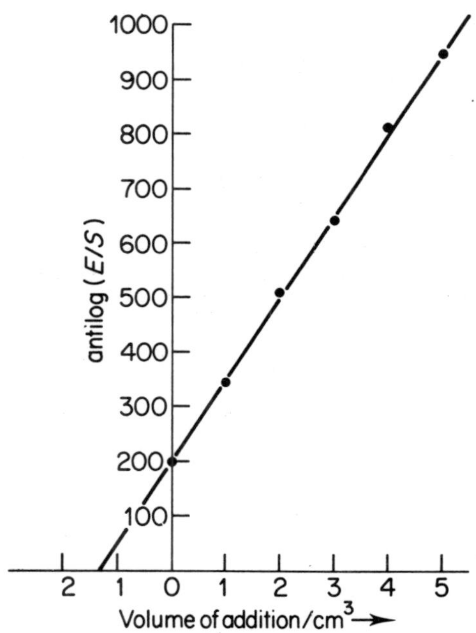

Fig. 2.6b. *Plot for multiple standard addition*

Applying Eq. 2.6f gives the concentration of the original sample as:

$$c = -(-1.34 \times 0.100)/100$$

$$= 1.34 \times 10^{-3}\ \text{mol dm}^{-3}$$

An alternative to calculating antilog(E/S) is to use a special form of graph paper called *Gran's Plot Paper*. This has one axis which when values of E(cell) are plotted directly on it converts them into antilog(E/S) values. The other axis is linear for plotting the volume. *Gran's* plot paper can be obtained in non-volume corrected and volume corrected forms.

The technique of multiple standard addition is used when the sample contains a high level of an interfering ion which is greater than that tolerated in direct measurement. After the standard additions, the concentration of the ion under investigation will be very much greater than that originally present, in which case it may swamp the interfering ion. As we extrapolate back from these interference free measurements, the influence of the interfering ion is very much reduced.

Multiple standard addition methods are also used when the ion under test is complexed, remember that ion selective electrodes only respond to the free ion. Whilst the comparative accuracy of addition and multiple addition techniques over direct reading methods varies from system to system, they are often more accurate. They are more time consuming, however, and the direct reading method is best retained when a large number of similar strength solutions are to be analysed. Addition methods gain favour for one off analyses when the time spent in calibration is not recovered by a large sample throughput.

Standard subtraction methods are not used as widely as the techniques previously discussed. They operate by adding a reagent which quantitatively removes some of the ion under investigation. This is particularly useful when it is not possible to make up standard solutions of the ion under test. The procedure is the same as for addition except that a minus sign appears in the concentration term of the Nernst equation.

Before subtraction:

$$E_1 = E^* \pm S \lg c \qquad (2.6g)$$

after subtraction:

$$E_2 = E^* \pm S \lg (c - x) \qquad (2.6h)$$

where x is the amount by which the concentration of the ion is reduced by reaction with the reagent. This can generally be calculated from a stoichiometric equation representing the reaction. Subtraction of Eq. 2.6g and 2.6h and rearrangement gives:

$$c = x/\{1 - \text{antilog}[(E_2 - E_1)/ \pm S]\} \qquad (2.6i)$$

SAQ 2.6b

A fluoride ion selective electrode was used to measure the concentration of F^- in a cup of tea. When immersed in a mixture of 25 cm^3 of tea and 25 cm^3 of ionic strength adjustor, the electrode gave a reading of 98 mV. When 2.0 cm^3 of a 100 ppm solution of F^- was added to this mixture, the measured emf dropped to 73 mV. Calculate the concentration of fluoride ions in the tea.

SAQ 2.6b

SAQ 2.6c

50 cm^3 of a solution of Cu^{2+} was analysed using a multiple standard addition method. When incremental amounts of a standard 0.100 mol dm^{-3} solution of Cu^{2+} were added to the sample, the following emf readings were obtained.

Volume of addition /cm^3	emf /mV
0	99.8
1.00	102.5
2.00	104.6
3.00	106.3
4.00	107.9

A blank solution gave an emf reading of 70.0 mV.

Estimate the concentration of the copper(II) ions in the original solution. (If Gran's plot paper is available use this.)

2.6.3. Potentiometric Titrations

In a normal volumetric titration such as:

$$AgNO_3 + NaCl = AgCl + NaNO_3$$

where aqueous silver nitrate is added to aqueous sodium chloride, we generally have two possibilities for detecting the equivalence point.

- We can detect the point when free chloride ions are no longer present in the solution, ie when all of the NaCl is used up; or

- we can detect when free silver ions first appear in solution, ie when $AgNO_3$ is in excess.

Conventional colour change indicators work on this principle of detecting the point of disappearance or appearance of one of the species involved in the titration reaction. The colour change then gives the end point rather than the equivalence point of the titration.

In potentiometric titrations we continuously monitor the activity of one species as it changes during the course of the titration. In the above example we could use the chloride ise to monitor the change in activity of the Cl^- ion. The cell emf is given by:

$$E(\text{cell}) = E^* - S \log a(\text{Cl}^-)$$

As the titration proceeds the activity of the chloride ion decreases. The term $\log a(\text{Cl}^-)$ becomes more negative (if you don't believe it, look up the value of the logarithm of 0.1 and 0.00001 on your calculator) and $[- S \log a(\text{Cl}^-)]$ becomes more positive. The measured emf would, therefore, increase during the course of the titration. This increase is not linear and a sigmoid shaped curve is obtained, as illustrated in Fig. 2.6c.

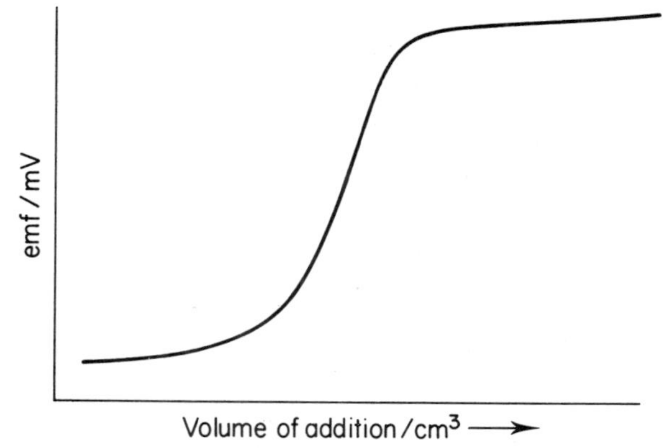

Fig. 2.6c. *Typical plot of emf against volume for a potentiometric titration*

The greatest change in emf occurs around the equivalence point. In fact the equivalence point itself is the point of maximum change in emf with volume added. This is termed the point of inflexion.

We now have a way of finding equivalence points of titrations without worrying about colour changes or overshooting in our titres. Further, potentiometric titrations are far more sensitive than normal colour change indicator titrations and we can titrate more dilute solutions.

The use of points of inflexion is only one method of obtaining equivalence points from emf/volume data obtained during the course of the titration. Other methods include

– plotting of the first derivative;

– plotting of the second derivative; and

– Gran's plot.

The first derivative method is based on the sigmoid shaped curve described in Fig. 2.6c. The slope of the curve at the start of the titration is very small – the line is almost horizontal. As the titration proceeds, then the slope increases to a maximum value at the equivalence point. Beyond the equivalence point the slope decreases to a very small value again.

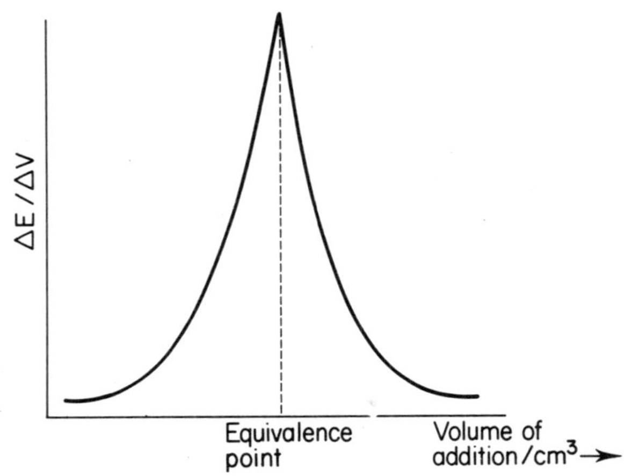

Fig. 2.6d. *Typical first derivative curve for a potentiometric titration*

We can represent this on a plot of slope against volume as illustrated in Fig. 2.6d. The equivalence point of the titration lies at the maximum of the peak. It is generally very much easier to estimate this maximum value than to find the point of inflexion. Let us see how we can plot the graph.

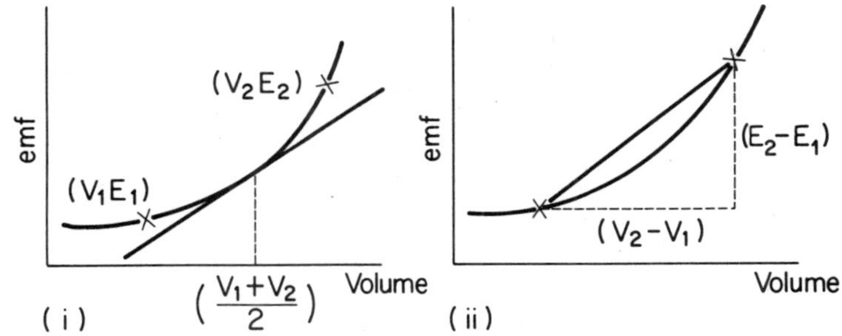

Fig. 2.6e. *Plots showing (i) the slope of the tangent at $(V_1 + V_2)/2)$ and (ii) calculation of the slope of the line joining two points. (The two lines are parallel)*

If we examine an enlarged portion of the original emf against volume plot, then we see the picture illustrated in Fig. 2.6e(*i*). The problem that we are faced with is to find the slope of a curve. It is easy to find the slope of a straight line joining the two points as shown in Fig. 2.6e(*ii*). But the line joining these points is parallel to the tangent to the curve half way between V_1 and V_2 ie at $(V_1 + V_2)/2$.

We can repeat this calculation for the tangent at $(V_2 + V_3)/2$, ie

$$\text{Slope} = \Delta E/\Delta V = (E_3 - E_2)/(V_3 - V_2)$$

We can now get a series of values for the slope of the sigmoid shaped curve at different volumes. We can then plot these points to give the first derivative as a function of volume.

Π The titration of Fe^{2+} solution with 0.1095 mol dm^{-3} Ce^{4+} solution was monitored potentiometrically using a platinum/reference electrode system. The following results were obtained.

Volume Ce^{4+} /cm^3	emf /mV
1.00	373
5.00	415
10.00	438
15.00	459
20.00	491
21.00	503
22.00	523
22.50	543
22.60	550
22.70	557
22.80	565
22.90	575
23.00	590
23.10	620
23.20	860
23.30	915
23.40	944
23.50	958
24.00	986
26.00	1067
30.00	1125

Using this data find the equivalence point by a first derivative method.

We must firstly calculate the slope and midway values for the volumes.

In the following table

$$\text{Slope} = \Delta E/\Delta V \text{ and } V' = (V_i + V_{i+1})/2.$$

Volume Ce^{4+} /cm^3	emf /mV	Slope /mV cm^{-3}	V' /cm^3
1.00	373	10.5	3.00
5.00	415	4.6	7.50
10.00	438	4.2	12.50
15.00	459	6.4	17.50
20.00	491	12	20.50
21.00	503	20	21.50
22.00	523	40	22.25
22.50	543	70	22.55
22.60	550	70	22.65
22.70	557	80	22.75
22.80	565	100	22.85
22.90	575	150	22.95
23.00	590	300	23.05
23.10	620	2400	23.15
23.20	860	550	23.25
23.30	915	290	23.35
23.40	944	140	23.45
23.50	958	56	23.75
24.00	986	40.5	25.00
26.00	1067	14.5	28.00
30.00	1125		

This graph is plotted in Fig. 2.6f, the equivalence point stands out very clearly at 23.15 cm^3.

The second derivative method is simply an extension of the first derivative method. If we plot a graph of the slope of the first derivative curve against volume we will get the shape shown in Fig. 2.6g.

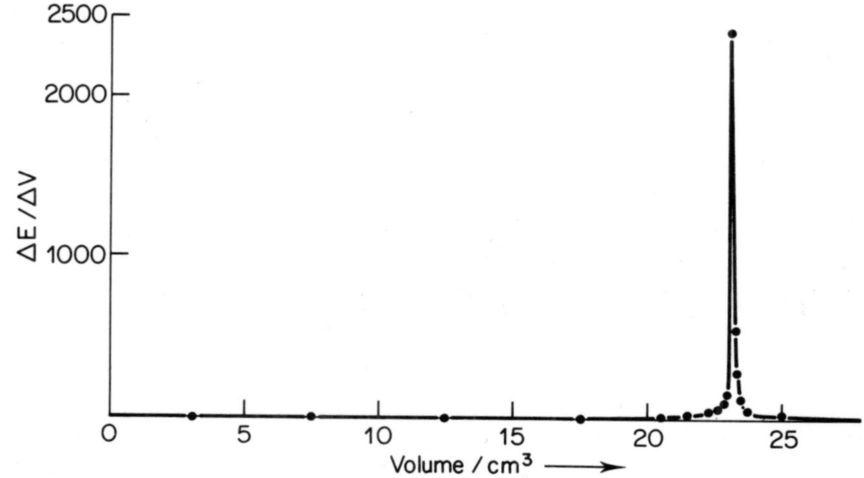

Fig. 2.6f. *First derivative plot showing maximum value at 23.15 cm*^{3}

Notice how the graph changes from a positive value to a negative value as the equivalence point is traversed. The point which the graph cuts the abcissa (the volume axis) is the point at which the slope is zero. This occurs at the maximum of the first derivative plot, ie the equivalence point.

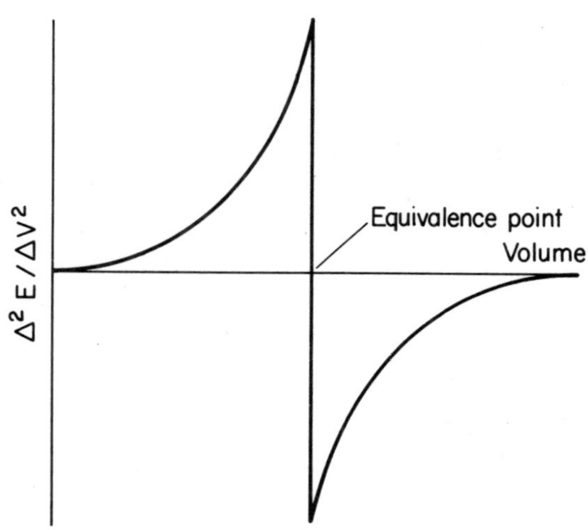

Fig. 2.6g. *Plot of second derivative against volume*

Π Using the data supplied in the previous question determine the equivalence point by the second derivative method.

All we have to do is to repeat the process. The slope of the initial sigmoid shaped graph was given by $\Delta E/\Delta V$. We write the slope of the second derivative as $\Delta^2 E/\Delta V^2$.

$\Delta E/\Delta V$	V'	$\Delta^2 E/\Delta V^2$	V''
10.5	3.00		
4.6	7.50		
4.2	12.50		
6.4	17.50		
12	20.50		
20	21.50		
40	22.25		
70	22.55	0	22.60
70	22.65	100	22.70
80	22.75	200	22.80
100	22.85	500	22.90
150	22.95	1500	23.00
300	23.05	21000	23.10
2400	23.15	−18500	23.20
550	23.25	−2600	23.30
290	23.35	−1500	23.40
140	23.45	−280	23.60
56	23.75	−12	24.50
40.5	25.00	−9	26.50
14.9	28.00		

Notice, I have taken a short cut in only working out the values over the regions of interest, ie around the equivalence point. The graph is illustrated in Fig. 2.6h.

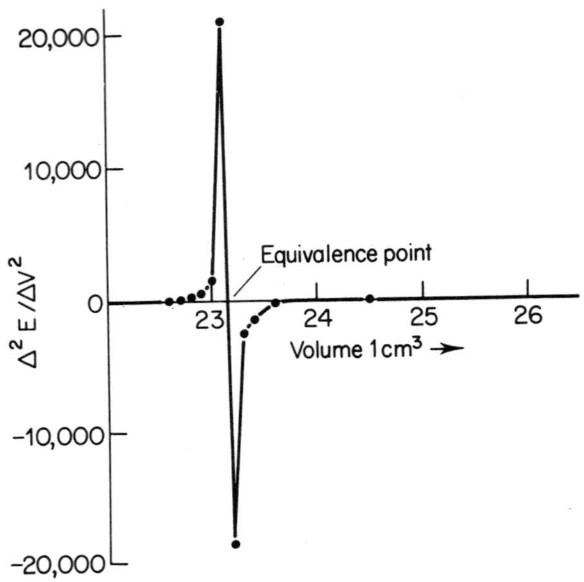

Fig. 2.6h. *Second derivative plot showing equivalence point at 23.15 cm³*

The third method on our list was Gran's Plot. Gunnar Gran was a scientist in a Swedish forest products laboratory, whose day to day job in the early 1950's involved endless potentiometric titrations. Having plotted numerous sigmoid shaped and derivative curves, he developed a method for determining equivalence points that was based on the extrapolation of a straight line plot.

Let us assume for a moment that we are following the concentration of a positively charged monovalent ion, i, which is part of the titration reaction. The Nernst equation will be:

$$E = E^* + S \log c_i$$

Rearranging this equation gives:

$$\log c_i = (E - E^*)/S$$

or

$$c_i = \text{antilog}[(E - E^*)/S]$$

$$= [\text{antilog}(-E^*/S)]\,[\text{antilog}(E/S)]$$

But, the first antilog term is a constant and, therefore:

$$c_i = \text{const.}\,[\text{antilog}(E/S)]$$

If the species, i, is disappearing during the course of the titration, eg Cl^- ions in a solution of NaCl when $AgNO_3$ is added, then it will do so linearly. Consequently, the antilog term will also reduce linearly and a plot of antilog (E/S) against volume of reagent added (V) should give a straight line whose intercept with the volume axis will be the equivalence point as shown in Fig. 2.6i.

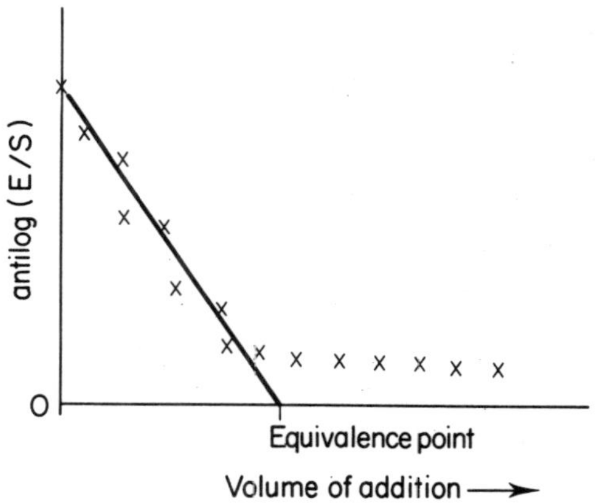

Fig. 2.6i. *Typical Gran's plot for a potentiometric titration*

Note that in the above titration, the first nine points lie on a reasonable straight line, whilst close to the equivalence point, there is considerable curvature. This is due to the limited solubility/dissociation of the precipitate/complex formed in the titration reaction. This is readily compensated for by the extrapolation of the straight line portion of the graph to its intercept on the abcissa.

Π What do you think the shape of the Gran's plot would be if the electrode monitored the titrant?

The plot would be the mirror image of the above with antilog (E/S) rising linearly after the equivalence point.

To further simplify the process, we can use Gran's Plot paper (available commercially from Orion Research Inc.), in place of calculating antilogs. This paper is available either as 'no volume correction' or '10 % volume corrected' sheets. With the latter it is necessary to adjust the dilution to 10 %, eg by adding 10 cm^3 to 100 cm^3. Clearly, you need to have some idea of the strength of the analyte solution to prepare a titrant of suitable concentration.

Π In the titration of 100 cm^3 of an approximately 10^{-2} mol dm^{-3} solution of Cl^- ions, what concentration of $AgNO_3$ must be used to keep the dilution to 10%?

We like to get our equivalence point about mid way through the titration. As we can add 10 cm^3 for a 10% dilution, we want the equivalence point to be at around 5 cm^3. Therefore:

$$5\ cm^3 \text{ of } x \text{ mol } dm^{-3}\ AgNO_3 = 100\ cm^3 \text{ of } 10^{-2} \text{ mol } dm^{-3}\ NaCl$$

As Ag^+ and Cl^- react in a 1 : 1 ratio, the $AgNO_3$ must be 20 times stronger than the NaCl.

Concentration of $AgNO_3$ = 0.2 mol dm^{-3}.

When we use Gran's plot paper, for potentiometric titration it is necessary to run a blank titration. We can then extrapolate this back to zero addition to give us an accurate location of the horizontal axis.

There are several advantages in using linear plots of this type. It is only necessary to obtain a few points to define the straight line. The points can be determined away from the equivalence point so that one reagent is in excess. This prevents dissociation and, as the concentrations are higher, the electrode responds quicker.

SAQ 2.6d

A student calibrated a Ca^{2+} ion selective electrode using two standard solutions at constant ionic strength and 298 K. He obtained the following results

Ca^{2+} concentration /mol dm^{-3}	emf /mV
1.00×10^{-3}	142
1.00×10^{-4}	113

What is the concentration of a test solution which gave an emf reading of 125 mV under the same conditions?

SAQ 2.6e

A 50.0 cm^3 sample of a lead(II) solution was analysed using a standard addition method with a lead(II) ion selective electrode. The initial potential was 302 mV, which rose to 350 mV on addition of 1.00 cm^3 of a 1.00 mol dm^{-3} solution of lead(II). Assuming that all measurements were made at constant ionic strength and 298 K, calculate the concentration of the original solution.

SAQ 2.6f

In the potentiometric titration of 100 cm^3 of a chloride solution with a 2.00×10^{-3} mol dm^{-3} solution of silver nitrate, the emf was measured using an Ag^+ ise-reference assembly. The following results were obtained.

Volume $AgNO_3$ /cm^3	emf/mV blank	titration
1.00	278	
1.50	287	
2.00	294	
2.50	299	262
3.00	304	273
3.50		282
4.00		288
4.50		294
5.00		297
5.50		300
6.00		304

Using the Gran's plot method determine the equivalence point of the titration and the concentration of the chloride solution. If you have access to 10% volume corrected Gran's plot paper use this and take the blank into consideration. If not plot antilog(E/S) against V (and ignore the blank readings).

Summary

Ion selective electrodes can be used to perform analyses.

Direct reading methods are quick and ideal for a large number of samples. Calibration plots using a number of standards can be established, a quicker, but less accurate, method of one or two point calibration is considered.

Incremental methods are ideal for a small number of samples or if the level of interference is high. Standard addition, multiple addition or standard subtraction variations are available. It is assumed that the electrode has a Nernstian slope.

The most accurate analytical method involving ion selective electrodes makes use of them as indicators in titrations. The technique of potentiometric titration is straightforward, though there are a number of ways in which the data may be handled; plot straight emf against volume curves, first derivative curves, second derivative curves or even use Gran's plots.

Objectives

You should now be able to:

- use the data from direct reading, incremental methods and potentiometric titrations to obtain the concentration of a test solution.

2.7. REFERENCE ELECTRODES

Overview

We have spent some time considering the indicator electrodes and their responses. It is now appropriate to consider the other half cell and examine how we construct some common reference electrodes.

Introduction

In the introduction to this Part we saw that galvanic cells were composed of two half cells. One of these consisted of the solution under investigation and the indicator electrode. The potential of this half cell varied in accordance with the Nernst equation and the activity of the relevant ion.

For the potential of the cell as a whole to reflect the indicator half cell potential it was necessary to keep the potential of the other half cell a constant. We called this second half cell the reference electrode. In this section we are going to consider some systems which are suitable for use as reference electrodes.

Π What are the important features to consider when constructing a reference electrode?

- The overriding consideration is that the potential of the electrode should remain constant, even if small amounts of current should pass through it, ie the potential of the electrode should not 'drift';
- the potential should be reproducible;
- the electrode should be easy to assemble;
- its potential should not change significantly with temperature;
- it would be very helpful, in the current climate of financial constraints, if it was cheap.

For any half cell, the electrode potential will be given by the now familiar, Nernst equation:

$$E(\text{cell}) = E^{\ominus} \pm S \log a_{\mathrm{i}}$$

If we keep the activity of i constant, then E(cell) should also remain constant. On the surface it seems easy to prepare a reference electrode by using a standard solution of a given ion. For example we could use a zinc rod dipping into a a 1.00 mol dm^{-3} solution of Zn^{2+}.

Π Whilst this looks simple enough a reference electrode prepared in this way would not satisfy all the conditions listed above. Can you suggest a reason why not?

If a small current flows through the half cell, an electrode reaction would occur. Depending on the direction of this current flow, either zinc would dissolve as zinc ions or vice versa. In either event, the activity of the zinc ions would change and with it the potential of our reference electrode. Further, it is relatively time consuming to prepare standard solutions.

We must now find a system which can either replace any ions that are lost or accept any ions that are produced by the electrode reaction. To achieve this we make a saturated solution of the zinc salt. Then if zinc ions are produced in the electrode reaction, more solid zinc salt will be formed. On the other hand, if zinc ions are deposited in the electrode reaction, more of the solid salt will dissolve to replace them. The activity of the zinc ions remains constant and, with it, the potential of the reference electrode.

The solubility of most zinc salts is relatively high and we would need to dissolve an awful lot of the salt. If we chose a sparingly soluble salt we could achieve the same effect more easily and often more cheaply. The three common sparingly soluble salts employed in reference electrodes are mercury(I)chloride (calomel), silver chloride and mercury(I)sulphate(VI). We will firstly see how the reference electrodes based on these three salts work and then examine the problems associated with each.

2.7.1. Saturated Calomel Electrode (sce)

Calomel is mercury(I)chloride, Hg_2Cl_2. The reference half-cell involving it may be represented:

$$KCl(satd) \mid Hg_2Cl_2(satd) \mid Hg$$

and the electrode reaction is:

$$Hg_2Cl_2(s) + 2e \rightleftharpoons 2\,Hg(s) + 2\,Cl^-$$

The electrode potential for this cell is +0.242 V against the standard hydrogen electrode (she) at 25 °C.

The form and shape of the sce varies from manufacturer to manufacturer and a typical example is illustrated in Fig. 2.7a. It consists of a central tube which contains the components of the electrode reaction, ie mercury, mercury(I) chloride and free chloride ions supplied by the KCl. The bridge solution fulfils two functions. Firstly it acts as a 'top up' to prevent the cotton wool plug from drying out. Secondly it acts as the salt bridge linking reference and indicator half cells.

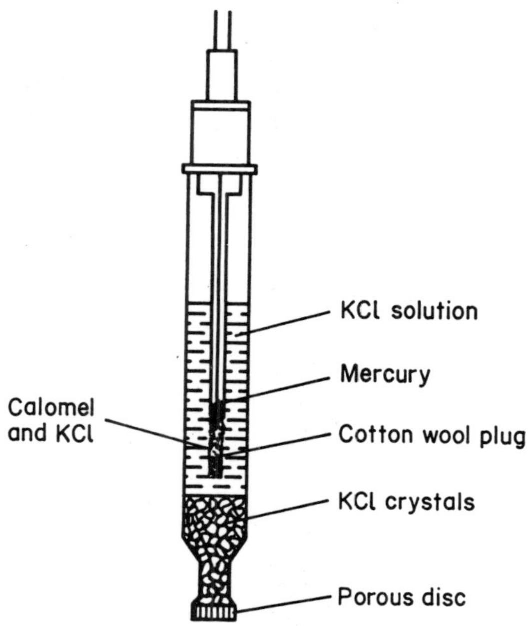

Fig. 2.7a. *Calomel electrode*

The porous divisions forming the boundaries at the liquid junctions are usually made from a ceramic material. This provides good electrical contact between the solution, but prevents a major interchange of the two liquids. There are other types of junction available, but most are declining in popularity.

2.7.2. Silver, Silver Chloride Electrode

This is the most common reference electrode in current use. It can be represented by:

$$KCl(satd),AgCl(satd) \mid Ag$$

and the half cell reaction is:

$$AgCl(s) + e \rightleftharpoons Ag(s) + Cl^-$$

The electrode potential of this electrode is +0.2046 V versus the she at 25 °C. The construction of the electrode is similar, in some respects, to the sce, and is illustrated in Fig. 2.7b. The silver wire, coated by a layer of silver chloride, dips directly into a solution of KCl, to which a few drops of silver nitrate(V) solution have been added. This solution is then saturated with respect to AgCl. This solution of KCl acts both as the internal solution and as the salt bridge.

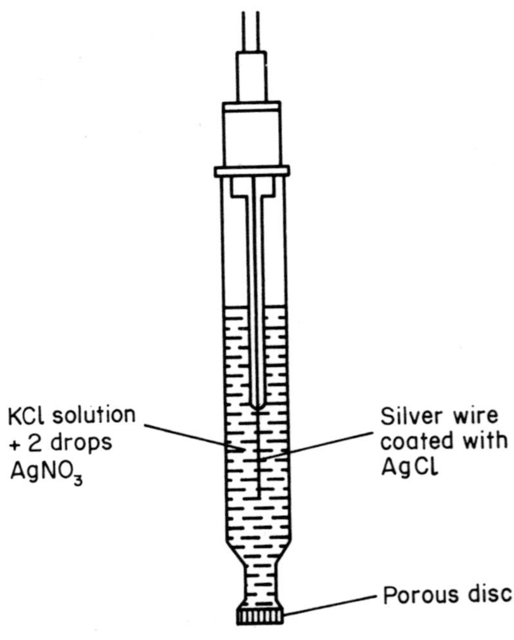

Fig. 2.7b. *Silver, silver chloride electrode*

2.7.3. Mercury, Mercury(I)sulphate Electrode

This electrode resembles the sce very closely. The half cell may be represented by:

$$K_2SO_4(satd) \mid Hg_2SO_4 \mid Hg$$

and the electrode reaction is:

$$Hg_2SO_4(s) + 2e \rightleftharpoons 2\,Hg(s) + SO_4^{2-}$$

This electrode has a potential of +0.412 V relative to the she at 22 °C.

Mercury(I)sulphate(VI) undergoes hydrolysis and a yellow precipitate of the basic salt is often seen in these electrodes. Fortunately, this reaction does not appear to affect the potential of the electrode at normal temperatures.

2.7.4. Factors Affecting Effectiveness of Electrodes

We can compare the effectiveness of the electrodes under a number of headings.

(*a*) *Temperature*

The temperature coefficients of the three electrodes are tabulated:

Electrode	Temp coeff/mV K^{-1}
sce	+0.19
Ag,AgCl	+0.09
Hg,Hg_2SO_4	+0.13

These values are due to the temperature dependence of the Nernst equation itself and also due to the variation in solubility of the appropriate salts with temperature. Whilst these values apply to the electrodes as described, they will be different if the concentration of the KCl or K_2SO_4 solutions are altered. In any event, the silver/silver chloride electrode is generally the most stable to temperature change.

(*b*) *Contamination of the internal electrode solution*

One of the major reasons for drift in reference electrodes is that interfering species can diffuse across the ceramic disc from the solution under investigation into the internal solution. This is especially true of the Ag,AgCl electrode where the internal solution and the bridge solution are one and the same. Common reagents which interfere with the electrode are redox compounds, sulphide, bromide, and cyanide ions. To minimise the problem we can use an electrode with a double junction as illustrated in Fig. 2.7c. This additional solution prevents the passage of interfering species from the solution under test to the internal solution. It does, unfortunately, present us with a further junction between two liquids. In potentiometry we attempt to keep the number of liquid junctions to a minimum as each creates a liquid junction potential (ljp) which will form part of the measured emf:

$$E(\text{cell}) = E(\text{right}) - E(\text{left}) + E(\text{ljp})$$

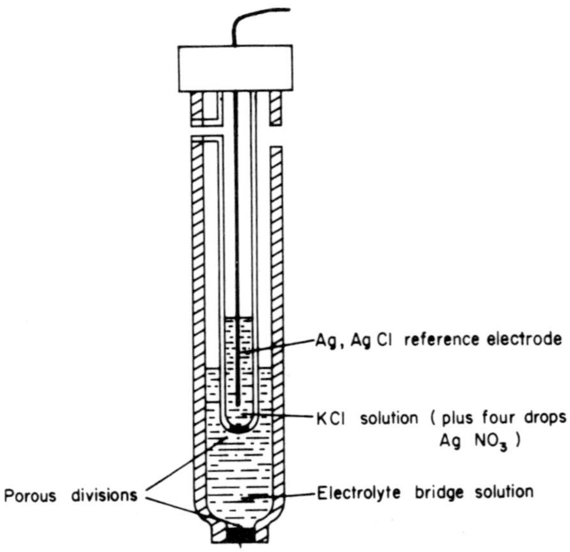

Fig. 2.7c. *Double junction reference electrode system*

Each ljp is a function of the relative mobilities of the negative and positive ions across the interface. KCl is a popular choice as a bridge solution due to the almost equal mobility values of the two ions. It is also advantageous to make the bridge solution very much more concentrated than that of the sample solution. This is because the ljp is more a function of the concentrated solution than the dilute solution. Therefore, if the bridge solution is very much more concentrated than the test and standard solutions, the magnitude of the ljp is virtually constant for all measurements. The value of this ljp is then incorporated into the constant potential of the reference electrode.

With double junction electrodes it is common to use a concentrated solution of KCl as the internal bridge solution with a solution of the ionic strength adjustor in the external bridge. In this way, the internal ljp is a constant value determined largely by the KCl concentration. The external 1jp will be negligible as the solutions either side of the boundary are similar in chemical composition and strength.

Whilst we can overcome most of the problems associated with ljps, the added complication of double junction reference electrodes is best avoided.

(*c*) *Contamination of the solution under investigation*

The passage of species across the ceramic membrane of reference electrodes is not a one way process. The internal solution can leak outwards, particularly if the level of the internal solution is higher than the external solution. Typical leakage lies between 0.01 and 0.1 cm^3 for every 5 cm difference in the levels. This leakage is particularly important when we analyse solutions of Cl^- or use ises that are affected by Cl^-. In such cases, whilst double junction reference electrodes can prevent transfer, it is better to avoid the extra complication and use the mercury,mercury(I)sulphate(VI) electrode.

Other examples where care must be taken are when one species reacts with a species across the junction. For example, if an sce is used when the solution contains Pb^{2+} ions, the Cl^- ions diffuse outwards and the Pb^{2+} inwards. They will meet as they pass through the ceramic division. A precipitate of $PbCl_2$ will form, partially blocking the membrane. This causes an increase in resistance which affects the measured emf and the elecrtrode will drift. To overcome this, it is necessary to use a double junction electrode.

(d) Ease of construction

Π Examine the diagrams of the above electrodes and comment on which you think is the easiest to make?

The silver, silver chloride electrode has fewer components and is considerably easier and cheaper to prepare. It has the further advantage that silver and silver chloride are not as harmful as mercury and its salts.

There is little to choose between preparing the sce and the mercury,mercury(I)sulphate(VI) electrode, though the latter is considered by some to be slightly easier.

(e) Cost

This can fluctuate with the relative prices of silver and mercury. The Ag,AgCl is generally slightly cheaper by virtue of its simplicity.

(f) Size

In the construction of microelectrodes or when a reference electrode needs to be combined within the body of an indicator electrode, size is important. In the majority of such applications the Ag,AgCl electrode is used because the bridge and internal solutions are one and the same.

(g) Stability

All three electrodes are extremely stable when stored in solutions of KCl or K_2SO_4 as appropriate. Stabilities of the order of ± 0.02 mV over several months have been quoted. This stability decreases with increasing temperature. Calomel disproportionates above 70 °C and excessive drift occurs. With mercury(I)sulphate(VI) hydrolysis becomes too much of a problem above 60 °C and this electrode drifts hopelessly above this temperature. The limiting value for Ag,AgCl electrodes is 125 °C, though formation of complexes such as $AgCl_3^{2-}$ causes drift below this value.

Stabilities also decrease in the presence of interfering agents in the external solution. The Ag,AgCl electrode is the worse culprit in this respect.

SAQ 2.7a

Select reference electrodes for the following applications, giving reasons for your choice.

(*i*) For a combined glass electrode to measure pH.

(*ii*) For use with a Cl^- ise at 298 K.

(*iii*) For use with a potassium ise to analyse sea water.

(*iv*) For use with an F^- ise to examine drinking water.

SAQ 2.7a

Summary

The principles involved in selecting and preparing reference electrodes have been discussed. The three reference electrodes: calomel; silver, silver choloride and mercury, mercury(I)sulphate, have been described in detail and we have considered the factors affecting their use.

Objectives

You should now be able to:

- list the requirements of a reference electrode;
- describe some common reference electrodes;
- state how a double junction can prevent sample and reference electrode contamination;
- list the main factors to be considered when selecting a reference electrode.

2.8. CHOICE OF EXPERIMENTAL REQUIRMENTS FOR AN ANALYTICAL PROCEDURE

Overview

When we examine a potentiometric method for the analysis of a given analyte, there are a large number of factors that we must consider.

2.8.1. The Electrode

Π Can you think of some factors you would need to consider when selecting an electrode?

Here are some of the factors to consider.

- Is there an electrode available to perform the analysis directly? If not can you determine it indirectly using another electrode?

- Does the electrode exhibit Nernstian behaviour over the range you wish to measure?

- Are there any interfering ions that will affect the readings? If the selectivity coefficient is large, do you need to remove them or can you compensate by choice of method, eg multiple standard addition?

- Does the response time of the electrode allow you to perform the number of analyses required in the time available?

- Will the electrode be affected by the conditions under which it operates, eg temperature, light?

- How stable is the electrode over a long period?

- Can you afford it and how long will it last?

- Is the electrode robust enough for the environment of the analysis?

2.8.2. Choice of Method

Π Can you list some factors that would affect your choice of method?

There are many factors to consider – if you thought of most of the following you're doing well.

- Does the electrode sense the substance directly? If not an indirect method such as titration would have to be used.

- How many samples do you need to analyse? If it's a one-off analysis, you'd probably be better off using an incremental method or a titration. If you have a large number of samples, then a calibration graph followed by direct reading is quicker.

- Do you need to use a multiple standard addition technique to overcome interference?

- Do you need to recalibrate the electrode often because of drift. If so, will an incremental method be quicker than direct reading even for large number of samples?

- To what degree of accuracy do you need your answer? A potentiometric titration is generally more accurate than direct reading method. The standard addition is usually in between as far as accuracy is concerned.

- How accurately do you know the slope of the electrode response? If you have to prepare a calibration curve to find out, you may as well use the direct reading method, even if only a few samples are involved.

- Have you got a good standard to calibrate the electrode? If not, then titration may be the only method available.

You can see that some of these points appear under heading of choice of electrode. The truth of the matter is that in devising the method of analysis, we have to optimise all parameters.

2.8.3. Choice of Ionic Strength Adjustor

Ionic strength adjustors often have to perform a few extra functions in addition to giving a constant ionic strength. Consider the isa that is used with the fluoride (F^-) electrode. This is called a total ionic strength adjustor buffer or tisab for short. It is prepared from

- sodium chloride to give a constant ionic strength;
- a complexing agent 1,2-diaminocyclohexanetetraacetic acid (cdta) which removes aluminium and other metal ions. This is because aluminium also complexes with fluoride and its presence would remove F^-; remember that ion selective electrodes respond only to the free ion.
- a buffer reagent to ensure that the electrode operates in its optimum working range.

2.8.4. Choice of Reference Electrode

Here we must ensure that we have no contamination by ions moving across the boundary, no drift and an appropriate temperature response. We are also looking for electrodes that are cheap and easy to prepare and robust in service.

SAQ 2.8a

Identify the factors which you think are important in selecting:

(*i*) an indicator electrode;
(*ii*) a reference electrode and
(*iii*) a method,

for the routine potentiometric analysis of drinking water for calcium and magnesium ions (ie water hardness).

SAQ 2.8a

Summary

In this section we have discussed the factors that need to be considered in chosing a method for potentiometric analysis.

Objectives

You should now be able to:

- list the important factors in selecting an indicator and reference electrode, the choice of method and the ionic strength adjustor.

3. Method of Operation of Ion-selective Electrodes

Overview

In this part of the unit we will see how the different types of ion selective electrode work. We will start with a glass electrode, which is one of the most common. We will then look at other types of electrode such as solid state, ion exchange, enzyme and gas sensing probes. As each type of electrode is examined, we will be able to see how many of the factors described in the previous part, such as limit of measurement, are related to the construction of each type of electrode.

3.1. THE GLASS ELECTRODE

The major use of glass electrodes in the laboratory is the measurement of pH. Whilst there are other electrode systems which can be used for pH measurement, glass is by far the most convenient. However, pH measurement is only one application of glass electrodes and, with some modification, they can be used to analyse many other species such as Na^+, K^+ and NH_4^+. In this section we will deal with the normal pH glass electrode first and describe the modifications later.

3.1.1. Theory of the pH Electrode

The typical construction of a glass electrode is shown in Fig. 3.1a.

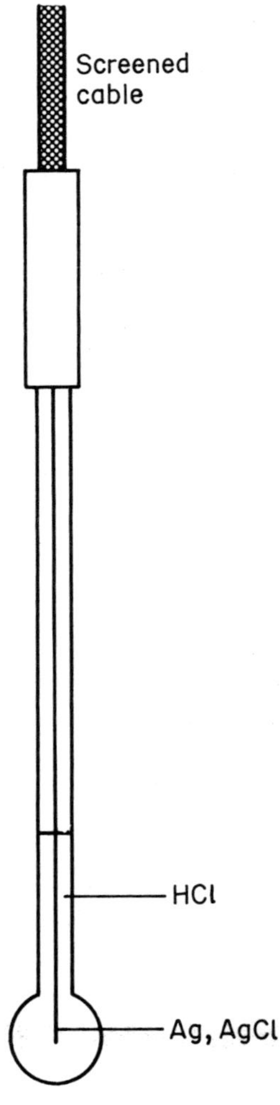

Fig. 3.1a. *Typical glass (pH) electrode*

A thin-walled glass membrane, which makes the electrode very delicate, encloses an internal standard solution, which is usually aqueous HCl of concentration around 1.0 mol dm^{-3}. Into the solution dips a rod of silver covered with silver chloride, which acts as an internal reference electrode. The overall galvanic cell, when the glass electrode is used with an external reference electrode, can be represented by:

Reference electrode (internal)	H+ (internal)	Glass membrane	H+ (external)	Reference electrode (external)
Glass electrode			analyte	Reference electrode

The key to the selectivity of the electrode is the glass membrane. The glass chosen consists of chemically bonded Na_2O and SiO_2 and is low in Al_2O_3 and B_2O_3. The surface layers of the glass consist of fixed silicate groups associated with sodium ions ($-SiO^-Na^+$) as illustrated in Fig. 3.1b.

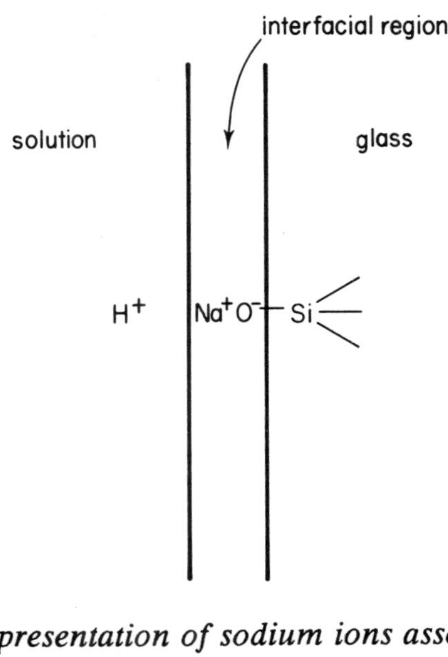

Fig. 3.1b. *Representation of sodium ions associated with fixed silicate groups at the surface of the glass electrode*

When this electrode is first soaked in water, the sodium ions ion-exchange with the solvated protons in water

$$\underset{\text{solid}}{-\text{SiO}^-\text{Na}^+} + \underset{\text{solution}}{\text{H}^+} \rightleftharpoons \underset{\text{solid}}{-\text{SiO}^-\text{H}^+} + \underset{\text{solution}}{\text{Na}^+}$$

The surface is now described as hydrated. The glass membrane will have an inner and outer hydrated layer, Fig. 3.1c. In these hydrated layers (typically 10^{-5} to 10^{-4} mm thick) the anion sites are covalently bonded to the bulk of the glass and are fixed. The cations (H^+), on the other hand, are not fixed leaving them free to exchange with the external solution or with sodium ions in the body of the glass.

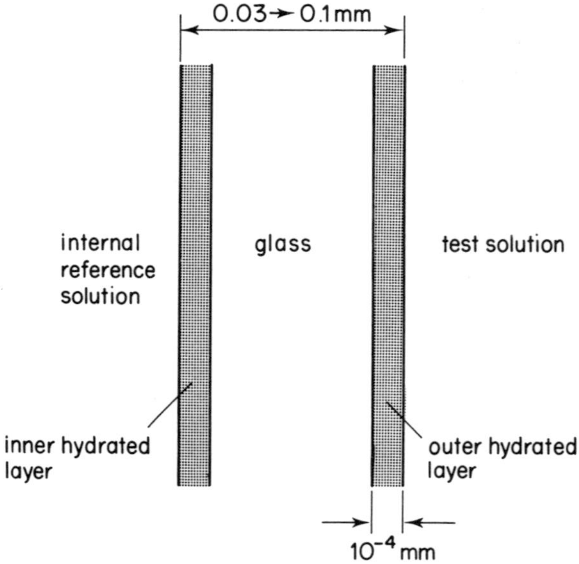

Fig. 3.1c. *Hydrated surface layers of glass membrane*

When the electrode is placed in a solution of unknown pH, the activity of the H^+ ions in the test solution is likely to be different to the activity of the H^+ ions in the hydrated layer. This sets up a potential difference between the solution and the surface of the membrane. This is termed a boundary potential and its magnitude will be determined by the difference in the activities. A similar boundary potential will exist at the inner surface of the membrane due to the hydration of that surface by the aqueous HCl.

The glass membrane, therefore, has two boundary potentials which can be represented by:

Outer solution	+	−	Glass membrane	−	+	Inner solution

The solution is represented as positive and the hydrated layer as negative. This would be true if the activity of H^+ was greater in the solution than in the hydrated layer. If the activity of H^+ in the hydrated layer was greater than in the solution, the signs would be reversed.

Π The magnitude of each boundary potential will be different. What factor(s) affect the magnitude of the overall potential across the membrane?

We can write the boundary potential for the outer surface:

$$\text{boundary potential, } E_b \propto (a - a_h)$$

where a and a_h are the activities of the H^+ ions in the test solution and in the hydrated layer respectively. We can write a similar expression for the inner surface. The overall potential will be the algebraic sum of the inner and outer boundary potentials:

$$\text{overall potential} \propto E_b\text{ (inner)} - E_b\text{ (outer)}$$

$$\propto (a - a_h) - (a_i - a_h)$$

where a_i is the activity of the hydrogen ions in the inner reference solution. If we assume that the activities of the H^+ ions in the inner and outer hydrated layers are the same, they will cancel out, thus:

$$\text{overall potential} \propto (a - a_i)$$

But the activity of the H^+ ions in the inner solution is a constant and the expression reduces to:

$$\text{overall potential} \propto a$$

Π We described the function relating activity to cell emf in Part 2. Write an expression for the emf of a cell containing a glass and Ag,AgCl reference electrode at 298 K.

The function relating the potential to the activity is the Nernst equation (Eq. 2.1a). Thus, for the glass electrode at 298 K we have:

$$E(\text{cell}) = E^* + 0.0591 \log a(\text{H}^+) \tag{3.1a}$$

where E^* includes the standard electrode potential of the glass electrode, the potential of the reference electrode and liquid junction potentials between the reference electrode and the solution. We define pH as:

$$\text{pH} = -\log a(\text{H}^+)$$

thus, substituting into Eq. 3.1a gives:

$$E(\text{cell}) = E^* - 0.0591 \text{ pH} \tag{3.1b}$$

or

$$E(\text{cell}) \propto \text{pH}$$

Whilst E^* includes the three potentials listed above, it also needs to include a fourth term, called the assymetry potential. This is a small potential which exists across the membrane, even when the inner reference solution and the test solution are identical (ie even when $a = a_i$).

Π Can you suggest any reasons why the assymetry potential exists?

There are a number of reasons.

- In the above derivation, we have assumed that the degree of hydration of the inner and outer surfaces of the glass is identical. In forming the glass membrane, the outside is subjected to a flame. This can result in the outer surface having a different structure to the inner. Consequently, there may be a different number of sites for the H^+ ions to exchange and the activity of H^+ in the inner and outer hydrated layers may be different.

However, the difference between them should be constant and it does not affect the fact that:

$$\text{overall potential} \propto a$$

It does however affect the value of E^*.

— The outer surface is also subject to mechanical and chemical attack whilst in use. This can result in a change in degree of hydration and the assymetry potential can change with electrode use. However, this change will be relatively slow and the assymetry potential due to it can be considered constant over a short period.

The degree of hydration of the outer surface will also change if the electrode is allowed to dry out, eg if it was left in air. It is, therefore, very important that the electrode be stored either in water or in damp cotton wool.

— If the glass is of non-uniform composition small potential differences can be set up due to areas of high and low Na^+ content.

— In the same way, if the glass contains impurities, then they can also set up small potential differences.

It is impossible to predict the magnitude of the assymetry potential and it is necessary to calibrate each electrode with buffer solutions. As the magnitude of the assymetry potential can change with electrode use, it is necessary to perform this calibration at regular intervals – at least once a day.

The glass electrode has a very high resistance (between 1 and 100 Mohm) and, consequently, it is vitally important to use an appropriate voltmeter. In practice high impedance voltmeters are used and are termed pH meters. The voltage scale is calibrated in pH units so that each decrease in 0.0591 V in E(cell) corresponds to 1 pH unit at 25 °C. This value will change with temperature and pH meters have a temperature control. This may either be set manually or by an automatic probe which dips into the solution.

SAQ 3.1a Sketch and label a conventional glass electrode.

SAQ 3.1b

Which of the following descriptions best describes the term 'boundary potential'?

(*i*) The effect of the hydration of the surface layers of the glass.

(*ii*) The effect of the difference in activity between the H^+ ions in the test solution and the hydrated layer.

(*iii*) The effect of the difference in the structure of the inner and outer glass surfaces.

(*iv*) The effect of the difference between the activity of the H^+ ions in the test and reference solutions.

SAQ 3.1c

Which of the following circumstances would result in a change in the assymetry potential (more than one of the list may contribute)?

(*i*) The electrode has been used to measure the pH of solutions in which the activity of the H^+ ions in the test solution and the inner reference solution are different.

(*ii*) The electrode has not been buffered for over a day.

(*iii*) The electrode has been stored in a cardboard box in a cupboard.

(*iv*) The electrode has been used to measure the pH of solutions containing sodium hydroxide.

3.1.2. Errors in the Measurement of pH

The glass electrode generally exhibits a Nernstian response over the bulk of the pH range. However, its behaviour becomes non-Nernstian at the extremes of pH giving two types of error, referred to as alkaline and acid error. We will now examine the reasons for this behaviour.

As the name suggests, alkaline error occurs at high pH values – above 9 or 10 pH units. The reason for the error is that whilst the glass membrane is selective to H^+, it also responds to other ions. This will be particularly significant if the activity of the other ion is high relative to the activity of the H^+.

Π Can you suggest some other ions which the glass membrane might respond to?

I hope you thought of Na^+ or K^+ or similar cations. If we re-examine the equilibrium expression for the hydration of the electrode

$$-SiO^-Na^+ + H^+ \rightleftharpoons -SiO^-H^+ + Na^+$$

We can then use the le Chatelier Principle to predict that if the concentration of Na^+ is high and that of H^+ is low, the equilibrium will move to the left, reducing the degree of hydration. This, in turn, will affect the boundary potential due to the H^+ ions. It will also set up a boundary potential due to the difference in the activities of the Na^+ ions in the test solution and the surface layer. This second boundary potential will have the same sign as that for H^+. The electrode then behaves as if there are more H^+ ions present than there actually are. This, in turn, produces a more acid or lower value for the pH.

Other cations will reduce the hydration of the surface to greater or lesser extents according to the general equation:

$$-SiO^-H^+ + M^+ \rightleftharpoons -SiO^-M^+ + H^+$$

Where M^+ is any monovalent cation. The larger the value for the equilibrium constant for this equilibrium, the greater the error. Note that the substitution is not restricted to monovalent ions, though the equilibrium constant will be relatively small, even for divalent cations. Fig. 3.1d shows the error produced by different cations at different concentrations. The sodium ion gives by far the greatest error, followed by lithium and potassium.

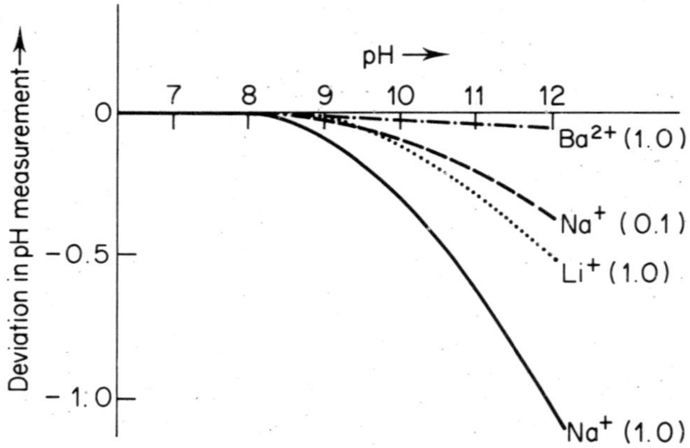

Fig. 3.1d. *Deviations in pH measurement using a standard glass electrode under alkaline conditions; the 'alkaline error' (Figures in parentheses refer to concentration of the ion in mol dm^{-3})*

Π Write the Nernst equation in a form that compensates for the effect of the Na^+ ion on the measured emf.

The Na^+ ion is an interfering ion and we use the modified form of the Nernst equation (Eq. 2.4b)

$$E(\text{cell}) = E^* + (RT/F) \ln [a(H^+) + k_{H,Na}a(Na^+)] \quad (3.1c)$$

where $k_{H,Na}$ is the selectivity coefficient of the electrode for H^+ over Na^+. It can be seen that the interfering ion effectively increases the H^+ activity giving the low pH reading we predicted earlier.

The fact that Na^+ has the largest selectivity coefficient of all the interfering ions is unfortunate. Not only do the majority of alkaline solutions contain sodium but also biological saline solutions and those in contact with sea water. Below pH 9, the high H^+ activity is sufficient to swamp the Na^+, but above this value it is advisable to modify the electrode slightly.

If the Na_2O in the glass is replaced by Li_2O, then $k_{H,Na}$ is very much less. Electrodes of this type are termed lithium-glass electrodes and they are used for measurement of high values of pH. The normal glass electrode is preferred below pH 10 because it has a faster response time and suffers less from drift.

At the opposite end of the pH scale, glass electrodes suffer from acid error as illustrated in Fig. 3.1e. This is sometimes described as 'water activity error', which gives some clue as to the reason. When writing the Nernst equation in this section we have ignored the activity of the water with which the solutions are made up. This is because that if the water is in a large excess, it behaves as a pure substance and we may take its activity to be unity. However, in highly acidic solutions, the activity of the water becomes less than unity because an appreciable amount is used in hydrating the protons. A similar effect is observed on addition of very large amounts of any dissolved salt as the concentration of free water is then reduced. The water can also be diluted and its activity reduced if a miscible non-aqueous solvent such as ethanol is added. The net effect in each case is that the measured pH will be too high.

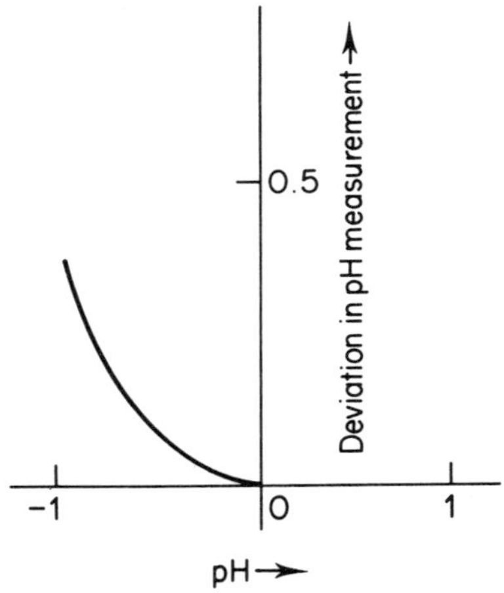

Fig. 3.1e. *Deviations in pH measurement using a standard glass (pH) electrode under acid conditions; the 'acid error'*

In all these cases, where the concentration of at least one species is very high, the liquid junction potential between the test solution and the reference electrode will also increase and contribute to the measured emf. This makes a further contribution to the measurement error.

SAQ 3.1d

Which of the following would cause an alkaline error?

(*i*) High activities of Na^+.

(*ii*) Low activities of H^+.

(*iii*) The presence of a non-aqueous solvent.

(*iv*) The glass membrane contains Li_2O.

SAQ 3.1d

3.1.3. Calibration and Accuracy of pH Measurement

We have seen in the previous sections that glass electrodes are subject to an assymetry potential and require regular calibration. The calibration process is relatively easy and requires the preparation of a standard pH solution, ie a buffer solution. Three factors need to be considered when calibrating a pH electrode.

Very few experimental chemists now prepare their own buffer solutions as chemical suppliers, (such as BDH) market a wide range. These either come in sachets or phials and are prepared by dissolving in a fixed quantity of distilled or demineralised water. Any buffer solution, however prepared, can only be quoted to ± 0.01 of a pH unit and this limits the accuracy of any pH measurement to a similar range. However, it is possible to measure the relative pH of two buffer solutions to ± 0.001 of a pH unit and, consequently, discriminate between two solutions with such a difference.

Another factor which limits the accuracy of pH measurements is the liquid junction potential between the reference electrode and the buffer or test solution. If an electrode is calibrated in a buffer solution of one composition and then used to measure the pH of a test solution of different composition, then the liquid junction potential will be different. This error, combined with the limits imposed by the buffer solution results in a total absolute accuracy of a pH measurement in the order of ± 0.02 pH units and a discrimination of ± 0.002 pH unit.

A third factor which needs to be considered is temperature. A change in temperature has many effects on the system. It alters the value of the Nernstian slope, it affects the activities of the ions (and, hence, the pH of the buffer) and it affects the liquid junction potentials. Commercial buffer solutions are generally provided with a table of pH values at different temperatures and it is advisable to calibrate the electrode at the temperature of the test solution and at approximately the pH value of the test solution. In this way our test measurements are performed under conditions which deviate very little from the known values at calibration.

Π Select appropriate conditions to calibrate an electrode which will be used to measure the pH of whole blood samples.

An example of such a calibration process occurs in the measurement of the pH of blood. It is advisable to calibrate the electrode at 38 °C with a buffer solution of pH around 7 – a phosphate buffer solution is suitable. This mirrors the conditions of the blood sample–body temperature and almost neutral pH.

SAQ 3.1e

List some factors which influence the accuracy of pH measurement.

SAQ 3.1e

3.1.4. Modified Glass Electrodes

We can make glass electrodes selective for ions other than H^+ by very simple modifications. In brief, these changes are in the composition of the glass and the internal filling solution. For example, a conventional glass electrode's selectivity can be summarised by:

$$H^+ >>> Na^+ > K^+, Rb^+, Cs^+, \ldots \gg Ca^{2+}.$$

By adding Al_2O_3 to the Na_2O and SiO_2 in the glass and changing the internal filling solution from HCl to NaCl, the electrode becomes much more responsive to Na^+. The response can now be summarised by

$$Ag^+ > H^+ > Na^+ \gg K^+, Li^+ \ldots > Ca^{2+}.$$

These early electrodes were referred to as NAS 11-18, which is an acronym of Na_2O, Al_2O_3 and SiO_2 coupled with their respective molar percentages – 11%, 18% and 71%. They have now been replaced by lithia based glass electrodes which use glasses containing Li_2O, Al_2O_3 and SiO_2.

Π From the above information, which conditions must apply when using sodium glass electrodes?

Both the NAS 11-18 and lithia based electrodes are more responsive to Ag^+ and H^+ ions. Interference from Ag^+ is not so common, but if it is present, it is easily removed by precipitation or complexation. H^+, on the other hand, is a common problem and its concentration needs to be kept constant in all test solutions. This is achieved using a buffer solution. When selecting buffer components, care must be exercised to ensure that they are sodium-free. A number of alternatives have been tried and one satisfactory buffer material is di-isopropylamine.

Another problem encountered with measuring Na^+ at very low concentration (below about 10^{-6} mol dm^{-3}) is that, if the solution is contained in a glass vessel, the sodium leaches out of the glass. This gives a false reading for each solution. We must, therefore, use metallic or plastic containers which are free of sodium.

Sodium electrodes are very useful and have a number of common applications, for example:

- measurement of sodium in biological fluids. Micro-electrodes now allow us to perform *in vivo* measurements, though in such cases the addition of an ionic strength adjustor buffer is not possible. This is not a problem as in the majority of biological applications we are interested in activity, rather than concentration, and we simply calibrate the electrode using activity standards;

- the determination of sodium in boiler feedwater. This involves the monitoring of distilled or demineralised water for Na^+ concentrations as low as 10^{-8} mol dm^{-3}. It is essential to remove all interfering ions and the measurements are performed at pH 11. The monitoring process is one that is repeated at regular intervals and is suitable for automation;

– estimation of sodium in natural waters, soil extracts and beer. At first, these applications appear to be very different, but the common feature is that the Na^+ concentration is relatively high. The normal method of analysis of such samples is flame photometry, though the sodium electrode compares favourably in terms of accuracy and precision. Further, the quoted response times lie between 14 and 40 s at 25 °C and the limit of measurement can be as low as 10^{-9} mol dm^{-3}, depending on conditions. The electrode has the advantage over flame photometry, that the sample solution does not require pre-treatment, such as filtering, before measurement.

These applications are only a very small selection and perhaps you can think of another sodium estimation that you have encountered.

Whilst the sodium electrode is the most common form of modified glass electrode, the NAS 27-4 electrode is also used for common cations. As the electrode name suggests, it is prepared using a glass containing 27% of Na_2O, 4% of Al_2O_3 and 69% of SiO_2. Its response to cations may be summarised by:

$$H^+ > Ag^+ > K^+ = NH_4^+ > Na^+ > Li^+ \ldots \gg Ca^{2+}.$$

Its use in the measurement of K^+ is now decreasing due to the development of another ise based on valinomycin. Whilst the latter is more selective, the NAS 27-4 glass electrode has certain advantages:

– it is cheaper;

– it has a lower detection limit;

– the membrane is relatively inert.

It, consequently, finds use where the concentration of interfering ions is low.

The NAS 27-4 electrode is as responsive to ammonia as it is to potassium. However, the ammonia gas sensing probe is more selective and the glass electrode finds little use in this application.

SAQ 3.1f

What effect would changing the composition of the glass from Na_2O/SiO_2 to $Li_2O/Al_2O_3/SiO_2$ have on a conventional glass electrode?

(*i*) Make the membrane more robust.

(*ii*) Make the electrode more selective towards H^+.

(*iii*) Make the electrode more selective towards Na^+.

(*iv*) Convert the electrode into an aluminium ise.

SAQ 3.1g

Which of the following ions would produce the greatest interference on a potassium glass electrode?

(*i*) Ag^{+};

(*ii*) NH_4^{+};

(*iii*) Na^{+};

(*iv*) Ca^{2+}.

Summary

The glass electrode is constructed using a thin walled glass membrane and is, consequently, very delicate. The conventional glass electrode responds to the activity of the H^+ ions in the test solution and can be used to measure pH. It is, therefore, commonly referred to as a pH electrode. When it is combined with the external reference electrode in one body, the name pH probe is often used. The electrode suffers from interference by sodium ions, particularly at high pH values, ie alkaline solutions.

Modifications in the composition of the glass can make the glass electrode selective to either sodium ions or potassium and ammonium ions. The sodium electrode is used in many applications as an alternative to flame photometry. The potassium/ammonium electrodes have now been largely superseded by other ion-selective electrodes using alternative membranes.

Objectives

You should now be able to:

- draw and label a conventional glass electrode;
- describe how a boundary potential is set up between the electrode surface and the test solution;
- write the Nernst equation for the glass electrode;
- describe the causes of assymetry potential and state how to allow for it;
- describe the causes of acid and alkaline error and list the sources and magnitude of errors associated with pH measurement;
- describe how conventional glass electrodes can be modified to act as selective electrodes for Na^+, K^+ and NH_4^+.

3.2. SOLID STATE MEMBRANE ELECTRODES

Overview

In the previous section, we saw that an electrode using a conventional glass membrane produces a Nernstian response to H^+ ions in solution. Changing the composition of the glass enables us to construct electrodes which have a Nernstian response to a number of different cations such as Na^+, K^+ and NH_4^+. In this section, we will take this a step further and change the membrane from glass to an ionic compound. This enables us to construct electrodes which have a Nernstian response to a number of different anions and cations such as F^-, Cl^- and Ag^+.

3.2.1. Construction of Solid State Membrane Electrodes

In many respects, the construction of solid state membrane electrodes mirrors that of the glass electrode. The basic principle is still

Test solution	Membrane	Internal reference

All solid state membrane electrodes use a robust body, generally of a polymer such as ABS, with a membrane fixed firmly at one end. Within this general picture there are three variations, illustrated in Fig. 3.2a, on how an internal reference is formed.

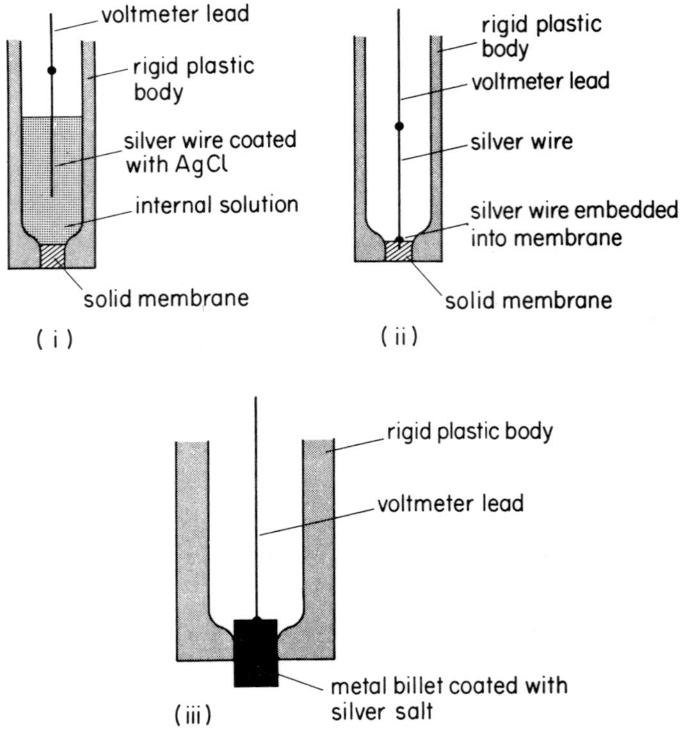

Fig. 3.2a. *Different methods of construction of solid state membrane ion selective electrodes, (i) internal reference solution; (ii) solid state configuration; (iii) coated metal*

(*i*) In the first variation, an internal solution and electrode form the internal reference. This is the same as in the glass electrode. One electrode constructed along these lines is that for F^-. This uses an internal solution containing KF, KCl and AgCl into which dips a Ag,AgCl electrode.

(*ii*) The second variation does away with the internal solution and connects the cable from the voltmeter directly to the membrane itself. The way in which this forms an internal reference is discussed later. This form of contact is common with membranes containing a silver salt. A silver wire is glued to the surface of the membrane using a epoxy resin containing finely divided silver to provide electrical contact. Electrodes of this type are sometimes referred to as 'all solid state'. Their main advantage is that there is no solution to evaporate or leak away.

(*iii*) The final variation involves coating the ionic compound onto the appropriate metal. The cable from the voltmeter is then connected directly to the metal as in the second type. In many respects, this type of electrode is very similar to the second variety. Both follow the general pattern

Test solution	Ionic Compound	Metal

The difference between (*ii*) and (*iii*) is the thickness of the ionic compound; in (*ii*), the membrane is relatively thick, whilst in (*iii*) it is merely a surface coating. The third type is much easier to construct and is very tough. However, as the coating is thin and porous, the solution can come into contact with the metal itself. This results in a loss of selectivity as the electrode is then responsive to redox changes in the solution.

The other variable in the construction of solid state membrane electrodes is how the membrane itself is made. The simplest is a homogeneous membrane consisting of a single crystal of the compound. This is machined into shape and fitted into the base of the electrode. If it is difficult to grow large crystals of the membrane material, then smaller grains of crystal can be compressed into a disc, in a similar fashion to preparing a KBr disc for infra-red analysis. This latter technique also allows mixture of ionic crystals to be used. For example, the solid state Pb^{2+} ion selective electrode uses a compressed disc of co-precipitated lead and silver sulphides.

Π Can you suggest one problem with using membranes made from a compressed disc.

If you have ever handled a KBr disc, you will realise how fragile it is. Whilst the compressed disc used in an ise is thicker, it also suffers from the problem of cracking whilst in use. To improve the mechanical strength, we can mix a non-active material with the ionic compound when making the disc. This additional material acts as a binder, giving the membrane greater strength. Typical binding materials are pvc, silicone rubber and poly(ethene). Membranes containing binders are referred to as heterogeneous. The amount of binding is kept to a minimum in order that sufficient contact occurs between adjacent grains of ionic crystal in the membrane. Fortunately, the presence of the binder does not affect the mode of operation of the electrode and we need consider it no further.

SAQ 3.2a

Sketch an all solid state ise using a silver sulphide membrane.

3.2.2. Method of Operation of Solid State Membrane Electrodes

In this section we will look at the basic theory behind the operation of solid state membrane electrodes. The method of operation of those electrodes with internal reference solutions is slightly different to the all solid state variety. The fluoride electrode will be used as an example of the former and those based on silver salts as an example of the latter.

The fluoride (F^-) ise utilises a single crystal of LaF_3 as the membrane. At both the inner and outer surfaces, an equilibrium is established between the F^- ions in the surface of the solid and the solution

$$F^-(aq) \rightleftharpoons F^-(s)$$

This is analogous to the establishment of the boundary potential in the glass electrode. Unlike the H^+ ions in the glass, the F^- ions in LaF_3 are relatively good conductors.

Π Suggest reasons why the F^- ions are able to carry charge through the LaF_3 membrane, whilst the H^+ ions do not carry charge through the glass.

The H^+ ions in the glass are only found in the outer hydrated layer. The positive sodium ions in the bulk of the glass are fairly tightly held in the lattice and are unable to move easily, Thus, glass is a poor conductor.

In the LaF_3 crystal, the F^- ions form an integral part of the entire structure. All ionic crystals contain defects and the F^- ions use these to move short distances. This is illustrated in Fig. 3.2b where one ion moves from its own lattice site to an adjacent vacancy. This leaves a vacancy at the original site, which can be filled by another ion and so on. The F^- ions are very much smaller than the La^{3+} ions and are, consequently, more mobile, virtually all the charge transfer in the crystal is due to F^-.

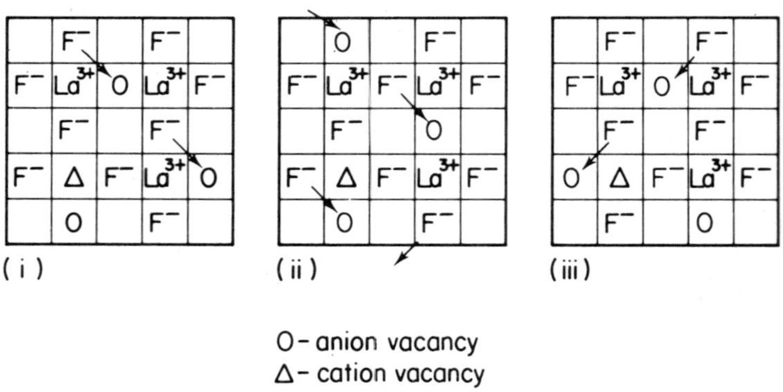

Fig. 3.2b. *Migration of fluoride ions via anion vacancies in the LaF_3 crystal. (The top surface of the crystal is in contact with a solution of higher F^- activity than the bottom surface)*

When the equilibrium between F^-(aq) and F^-(s) has been established at each interface, the activity of the F at the inner surface is likely to be different to that at the outer surface. ie

$$a(\mathrm{F}^-, \text{inner surface}) \neq a(\mathrm{F}^-, \text{outer surface})$$

This results in a diffusion potential being set up between the two surfaces, in a similar fashion to liquid junction potentials. As the ability to carry charge of the F^- and La^{3+} ions is different, a charge imbalance will be established. The magnitude of this imbalance is governed by the relative magnitudes of the inner reference and outer test solutions. As the former is a constant value, the diffusion potential is, thus, dependent on the activity of F^- in the test solution.

Π Write an equation to describe how the emf of a cell, containing a fluoride ise and an external reference, responds to the activity of F^- in a test solution at 298 K.

The response of an ise follows the Nernst equation (Eq. 2.2a). For an F^- ion, the sign of this equation will be negative and the magnitude of the charge on the ion will be 1. The slope of the electrode at 298 K is, therefore, 0.0591 V, if logarithms to the base 10 are used. The equation is then written as:

$$E(\text{cell}) = E^* - 0.0591 \log a(\mathrm{F}^-) \qquad (3.2a)$$

Where E^* incorporates the potential of the reference electrode and the standard potential of the ise.

Π Why is the electrode selective to F^-?

In an ionic crystal, the equilibrium

$$F^-(aq) \rightleftharpoons F^-(s)$$

exists at lattice sites on the surface. Only ions of a similar size, shape and charge to F^- can occupy these sites. As the combination of these three parameters is unique for F^-, the electrode is highly selective and suffers no interferences from ions such as Cl^-, Br^- etc.

The all solid state electrodes, based on insoluble silver salts, behave in a very slightly different way. The common salts used are silver halides (AgX) or silver sulphide (Ag_2S) and the electrode can be selective to either the cation (Ag^+) or the anion (X^- or S^{2-}).

The contact between the voltmeter cable and the inner surface of the crystal membrane is usually made in one of two ways. In the first method, a rod of pure silver is embedded in the crystal itself. In the second method, the voltmeter lead is glued to the surface using an epoxy resin. To provide the electrical contact necessary, the epoxy resin contains finely divided silver. Thus, in both cases, the inner surface is in contact with pure silver. An equilibrium will be established between the silver and the silver ions in the membrane:

$$Ag^+(s) + e \rightleftharpoons Ag(s)$$

The activity of the pure silver is a constant (unity) and this in turn fixes the activity of the silver ions at the inner surface. We have, therefore, established our inner reference as before.

When responding to silver ions in solution, the remainder of the electrode's behaviour is identical to that of the fluoride electrode, ie a boundary potential is established at the outer surface which, in turn, leads to a diffusion potential between the inner and outer surfaces.

Π Write an equation to describe how an all solid state membrane electrode and external reference electrode responds to silver ions in a test solution at 298 K.

As we have a monovalent positive ion, the Nernst equation is written as:

$$E(\text{cell}) = E^* + 0.0591 \log a(\text{Ag}^+) \quad (3.2b)$$

Π What part of this equation would be different if the electrode was constructed using an internal reference solution?

The boundary potential at the outer surface will still determine the response of the electrode and the equation will remain essentially the same. The only difference will be in the potential of the inner reference. This will affect the standard potential ($E^\ominus$) of the ise and, in turn, E^* of the cell. In the methods of analysis described in Part 2, no use was made of the E^* value and variation in construction will not affect our analysis, providing proper calibration is carried out.

We must now show that a solid state membrane, such as Ag_2S, gives a Nernstian response to the anion in solution.

Nernstian response to a sulphide anion at 298 K would be written as:

$$E(\text{cell}) = E^* - 0.0296 \log a(\text{S}^{2-}) \quad (3.2c)$$

We use a negative sign as we have a negative ion and the slope is half the value used for the silver ion because the sulphide ion is divalent (ie n = 2).

We must remember that the membrane still functions as an Ag^+ ise. However, at the outer surface, a solubility equilibrium is established according to:

$$K_{\text{sp}}(\text{Ag}_2\text{S}) = a^2(\text{Ag}^+)a(\text{S}^{2-}) \quad (3.2d)$$

Where $K_{sp}(Ag_2S)$ is the solubility product of silver sulphide. Some of the membrane will dissolve until the above equilibrium is established. Fortunately, Ag_2S is virtually insoluble and only very little will need to dissolve. After the equilibrium is achieved, there will be sulphide ions present from the solution and from the membrane. However, as the membrane is virtually insoluble the amount of S^{2-} obtained from it will be very small in comparison with that originally present in the test solution. The amount of S^{2-} present after equilibrium has been established can, therefore, be considered as being the same as that present in the original solution. Thus, in the solution equilibrium, the activity of the silver ions present is due to the membrane, whilst the activity of the sulphide ions reflects that of the original solution. We can rearrange Eq. 3.2d to give:

$$a^2\,(Ag^+) = K_{sp}\,(Ag_2S)/a(S^{2-})$$

or

$$a(Ag^+) = [K_{sp}(Ag_2S)/a(S^{2-})]^{\frac{1}{2}}$$

The ise responds to Ag^+ according to the equation:

$$E(\text{cell}) = \text{constant} + 0.0591 \log a(Ag^+)$$

Substituting for $a(Ag^+)$ gives:

$$E(\text{cell}) = \text{constant} + 0.0591 \log [K_{sp}(Ag_2S)/a(S^{2-})]^{\frac{1}{2}}$$

Splitting the log term gives:

$$E(\text{cell}) = \text{constant} + 0.0591 \log [K_{sp}(Ag_2S)]^{\frac{1}{2}} - 0.0591 \log [a(S^{2-})]^{\frac{1}{2}}$$

But as the solubility product of silver sulphide is a constant at a given temperature, the term $0.0591 \log [K_{sp}(Ag_2S)]^{\frac{1}{2}}$ is also constant. It can be incorporated into the constant already in the equation to give a new constant, which we can call E^*. Thus:

$$E(\text{cell}) = E^* - 0.0591 \log [a(S^{2-})]^{\frac{1}{2}}$$

but

$$\log[a(S^{2-})]^{\frac{1}{2}} = \tfrac{1}{2}\log[a(S^{2-})]$$

so that:

$$E(\text{cell}) = E^* - \tfrac{1}{2}\,0.0591\log[a(S^{2-})]$$

or

$$E(\text{cell}) = E^* - 0.0296\log a(S^{2-})$$

This is identical to Eq. 3.2c which we predicted for a sulphide ise at the start of this question. It shows that the electrode gives a Nernstain response to sulphide ions.

Note that the value of E^* for the electrode when it is used as an Ag^+ ise will be different to that when the electrode is used as an S^{2-} ise.

SAQ 3.2b

> List three factors which contribute towards the selectivity of solid state membranes.

SAQ 3.2c

Which of the following could be used as a solid-state membrane for an I^- ise?

(*i*) sodium iodide;

(*ii*) silver iodate;

(*iii*) iodoform;

(*iv*) silver iodide;

(*v*) lead(II)iodide.

3.2.3. Membranes Containing More than One Active Compound

We have seen that the silver ise can respond to the appropriate anion, in addition to Ag^+ itself. We can now extend this idea by forming membranes containing more than one active compound. For example, copper(II) sulphide (CuS) and silver sulphide (Ag_2S) can be co-precipitated from solution. The precipitate obtained is then compressed along with some binder into a disc, which acts as the membrane of the ise. The inner surface of the membrane is usually connected to a piece of silver wire, in the same way as normal all solid state electrodes. Let us now consider what happens when this electrode comes into contact with a solution of Cu^{2+} ions.

The electrode functions as an Ag^+ ise as discussed previously, though the equilibrium at the outer surface is slightly more complicated. Two solubility equilibria will be established in accordance with:

$$K_{sp}(Ag_2S) \quad = \quad a^2(Ag^+)a(S^{2-}) \qquad (3.2d)$$

and

$$K_{sp}(CuS) \quad = \quad a(Cu^{2+})a(S^{2-}) \qquad (3.2e)$$

The activity of the silver and sulphide ions will be due to the small amount of membrane that dissolves, whilst the activity of the copper ions will be that of the original solution. This is because the CuS membrane is virtually insoluble ($K_{sp} = 10^{-16}$ mol^2 dm^{-6}) and the amount of Cu^{2+} that dissolves is negligible when compared to that already present.

If we re-arrange Eq. 3.2e we get:

$$a(S^{2-}) = K_{sp}(CuS)/a(Cu^{2+})$$

If we substitute for the activity of the sulphide ions in Eq. 3.2d we have:

$$K_{sp}(Ag_2S) = [a^2(Ag^+)K_{sp}(CuS)]/a(Cu^{2+})$$

which on rearranging gives:

$$a^2(\mathrm{Ag^+}) = K_{sp}(\mathrm{Ag_2S})a(\mathrm{Cu^{2+}})/K_{sp}(\mathrm{CuS})$$

or

$$a(\mathrm{Ag^+}) = [K_{sp}(\mathrm{Ag_2S})a(\mathrm{Cu^{2+}})/K_{sp}(\mathrm{CuS})]^{\frac{1}{2}} \qquad (3.2f)$$

The Nernst equation for the silver ise, can be written as

$$E(\text{cell}) = \text{const} + 0.0591 \log a(\mathrm{Ag^+})$$

substituting for $a(\mathrm{Ag^+})$ from Eq. 3.2f gives:

$$E(\text{cell}) = \text{const} + 0.0591 \log [K_{sp}(\mathrm{Ag_2S})a(\mathrm{Cu^{2+}})/K_{sp}(\mathrm{CuS})]^{\frac{1}{2}}$$

which on splitting the log term becomes:

$$E(\text{cell}) = \text{const} + 0.0591 \log [K_{sp}(\mathrm{Ag_2S})/K_{sp}(\mathrm{CuS})]^{\frac{1}{2}} + 0.0591 \log a^{\frac{1}{2}}(\mathrm{Cu^{2+}}) \qquad (3.2g)$$

But the solubility products are constant at a given temperature and the term $0.0591 \log [K_{sp}(\mathrm{Ag_2S})/K_{sp}(\mathrm{CuS})]^{\frac{1}{2}}$ will also be a constant. We can incorporate this into the existing constant, to give a new constant E^*. Eq. 3.2g now becomes:

$$E(\text{cell}) = E^* = 0.0591 \log a^{\frac{1}{2}}(\mathrm{Cu^{2+}})$$

or

$$E(\text{cell}) = E^* + 0.0296 \log (\mathrm{Cu^{2+}})$$

This equation represents the Nernstian response to $\mathrm{Cu^{2+}}$ ions and demonstrates that the electrode acts as an ise for such ions.

We can extend this argument to any number of mixed sulphide electrodes and typical examples are given in Fig. 3.2c.

SAQ 3.2d

Silver sulphide is preferred to silver chloride when constructing a membrane for an Ag^+ ise because:

(*i*) it contains more silver ions;

(*ii*) its solubility product is lower;

(*iii*) chloride ions interfere;

(*iv*) it forms a stronger disc.

Selective ion	Membrane	Lower limit of measurement /mol dm^{-3}
F^-	LaF_3	10^{-7}
Cl^-	$AgCl$	10^{-5}
Br^-	$AgBr$	10^{-6}
I^-	AgI	10^{-8}
S^{2-}	Ag_2S	10^{-7}
SCN^-	$AgSCN$	10^{-6}
Ag^+	Ag_2S	10^{-8}
Hg^{2+}	HgS/Ag_2S	10^{-8}
Cu^{2+}	CuS/Ag_2S	10^{-9}
Cd^{2+}	CdS/Ag_2S	10^{-7}
Pb^{2+}	PbS/Ag_2S	10^{-7}
Bi^{3+}	Bi_2S_3/Ag_2S	10^{-11}

Fig. 3.2c. *Some examples of solid state membrane ion selective electrodes*

3.2.4. Selectivity

The selectivity of solid-state membrane electrodes revolves around reactions at the membrane surface. If we represent the ionic compound that forms the membrane as C^+A^-, and the interfering ion as either I^+ or J^-, two surface reactions are possible. For a cation ise, the positive ion will interfere:

$$C^+A^-(\text{s,membrane}) + I^+(\text{aq}) \rightleftharpoons I^+A^- \ (\text{s,surface}) + C^+(\text{aq})$$

whilst for the anion ise, the negative ion will have a similar effect:

$$C^+A^-(\text{s,membrane}) + J^-(\text{aq}) \rightleftharpoons C^+J^-(\text{s,surface}) + A^+(\text{aq})$$

The relative magnitude of the interference will depend upon the relative solubilities of the solids.

For example, LaF_3 is slightly less soluble than $La(OH)_3$. Thus, the equilibrium:

$$LaF_3(s,membrane) + 3\,OH^-(aq) \rightleftharpoons La(OH)_3(s,surface) + 3\,F^-(aq)$$

will lie over to the left hand side. The level of interference is relatively small, under normal operating conditions. However, when the concentration of F^- is very low and the concentration of OH^- is very high, the equilibrium will move to the right. Under these circumstances, the level of interference is significant.

The chloride ion is similar, in many respects, to the fluoride ion and could constitute a potential interfering ion. The chloride ion could enter into an equilibrium with the membrane:

$$LaF_3(s,membrane) + 3\,Cl^-(aq) \rightleftharpoons LaCl_3(s,surface) + 3\,F^-(aq)$$

Fortunately, the equilibrium constant for this reaction is so small, that even in the presence of a large excess of Cl^-, very little $LaCl_3$ would be formed. The selectivity constant, $k_{F,Cl}$, is very small and we can neglect the effect of Cl^- ions on the electrode. In fact, it is only the OH^- ion that interferes with the fluoride ise to any significant extent.

Π The solubility products of some silver salts are

AgCl	1.5×10^{-10} mol^2 dm^{-6}
AgBr	7.7×10^{-13} mol^2 dm^{-6}
AgI	0.9×10^{-16} mol^2 dm^{-6}

Predict the level of interference of Cl^- and I^- ions on a bromide ise using an AgBr membrane.

The two interfering reactions will be

$$AgBr(s,membrane) + Cl^-(aq) \rightleftharpoons AgCl(s,surface) + Br^-(aq)$$

and

$$AgBr(s,membrane) + I^-(aq) \rightleftharpoons AgI(s,surface) + Br^-(aq)$$

AgBr is less soluble than AgCl. Therefore, the first equilibrium will lie over to the left. Thus, chloride ions do not interfere significantly with the Br^- ise.

On the other hand, AgI is less soluble than AgBr. The second equilibrium will lie over to the right so that I^- interferes significantly with the determination of bromide ions.

We would expect the selectivity coefficient of the bromide ise for Cl^- ions to be small and that for I^- to be large. This is in fact, the case with $k_{Br,Cl} = 3 \times 10^{-3}$ and $k_{Br,I} = 5 \times 10^3$.

When we examine mixed compound membranes, represented as MS/Ag_2S, the sulphide of the ion to be determined (M^{2+}), is less soluble than Ag_2S. Interferences will, therefore, come from metal cations whose sulphides are less soluble than MS.

Whilst solubility or solubility product gives us an idea whether an ion has the potential to interfere, it is not the complete picture.

Π The solubility product of Ag_2S is approximately 10^{-49} mol^2 dm^{-6}. Why do not sulphide ions interfere strongly with electrodes based on silver halide membranes?

The equilibrium constant for

$$2\,AgX(s,membrane) + S^{2-}(aq) \rightleftharpoons Ag_2S(s,surface) + 2\,X^-(aq)$$

is certainly very large. However, the rate of the reaction from left to right is very slow Thus, over the time scale of a normal measurement, insufficient reaction occurs to create major problems. If the electrode was left in the solution over a longer period, the interference would manifest itself in the form of drift.

SAQ 3.2e

The solubilities (mol dm^{-3}) of some common sulphides are:

arsenic(III)	1×10^{-5}
cadmium(II)	2×10^{-6}
copper(II)	2×10^{-8}
germanium(II)	3×10^{-2}
iron(II)	1×10^{-4}
lead(II)	2×10^{-6}
manganese(II)	9×10^{-5}
tin(II)	2×10^{-7}

Predict the order which the following cations will interfere with the operation of a Pb^{2+} ise using a PbS/Ag_2S membrane.

As^{3+} Cd^{2+} Cu^{2+} Ge^{2+} Fe^{2+} Mn^{2+} Sn^{2+}

SAQ 3.2e

3.2.5. Response Times and Range

Solid-state membrane electrodes have the quickest response times of all ion selective electrodes. In well stirred solutions, the boundary potentials are established in fractions of a second. Under ideal conditions of small sample volumes and good stirring and with concentrations an order of magnitude or more greater than the limit of measurement, the electrodes will respond to hundredfold changes of concentration in milliseconds. Even in beakers containing larger volumes, response times in the order of seconds are obtained.

The range of concentrations over which the electrode responds, will vary from one membrane material to another. Let us firstly consider the upper limit of response. Both the LaF_3 and Ag_2S membranes are inert to the ion under investigation and will operate up to saturated solutions. The AgCl and other halide membranes will react with concentrated solutions of the anion to form haloargentate(I) complexes such as:

$$Ag^+Cl^-(s,membrane) + Cl^-(aq) \rightleftharpoons AgCl_2^-(aq)$$

and

$$Ag^+Cl^-(s,membrane) + 2\,Cl^-(aq) \rightleftharpoons AgCl_3^{2-}(aq)$$

This restricts the upper limit of measurement of such membranes to around 1 mol dm^{-3}. For the same reason, it is also advisable that the electrode is not stored in a concentrated solution of the anion when not in use.

Π The lower limit of measurement is also dependent on the nature of the membrane. Suggest which property of the membrane determines this lower limit.

In the vast majority of cases, the lower limit of measurement is determined by the solubility of the membrane. Consider the AgCl membrane. This dissolves according to:

$$Ag^+Cl^-(s) + H_2O\ (l) \rightarrow Ag^+(aq) + Cl^-(aq)$$

with the solubility product given by:

$$K_{sp}(AgCl) = a(Ag^+)a(Cl^-)$$

In our discussion on electrode response, we assumed that for the silver ions in solution after equilibrium had been attained:

$$a(Ag^+,\text{solution}) \gg a(Ag^+,\text{membrane})$$

Where the qualifications 'solution' and 'membrane' refer to the origin of the ions prior to equilibrium. As we attempt to measure progressively more dilute solutions, $a(Ag^+,\text{solution})$ becomes smaller and our assumption that:

$$a(Ag^+,\text{solution}) + a(Ag^+,\text{membrane}) = a(Ag^+,\text{solution})$$

does not hold and we will have deviations from Nernstian behaviour. As the solution becomes more dilute, we will reach a stage when:

$$a(Ag^+,\text{solution}) < a(Ag^+,\text{membrane})$$

The term $a(Ag^+$,membrane) will now dominate in the sum [$a(Ag^+$,solution) + $a(Ag^+$,membrane)]. However dilute the test solution, this sum cannot be less than $a(Ag^+$,membrane). This limits our response and we cannot measure halide ion activity lower than that provided by the membrane dissolving. Thus for the AgCl membrane, the limit of Nernstian response for Cl^- ions is about 10^{-4} mol dm^{-3}, whilst the lower limit of measurement is about 10^{-5} mol dm^{-3}. This is illustrated in Fig. 3.2d.

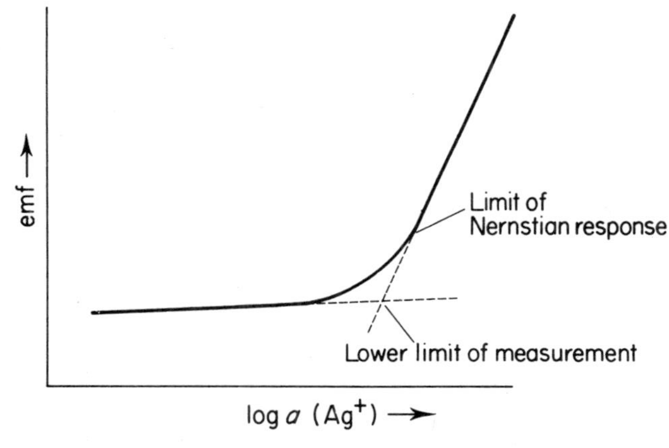

Fig. 3.2d. *Plot of log a(Ag^+) against cell potential for a AgCl solid-state membrane/reference electrode cell*

A similar argument can be used for other solid state membrane electrodes. If the surface of the electrode becomes contaminated by another substance, as happens when an interfering ion is present, the limit of measurement will not be as low. We have seen an example of this with the fluoride electrode. A coating of $La(OH)_3$ forms in extremely dilute solutions and this increases the response time so much, that the electrode cannot be used.

In brief, unless the membrane reacts with the test ion in the solution, the upper limit of measurement is a saturated solution. At the other end of the scale, the lower limit of measurement is a function of the solubility of the membrane.

SAQ 3.2f

Are the following statements true or false when applied to solid-state membrane electrodes?

(*i*) They have relatively fast response times.

(*ii*) They should be stored in concentrated solutions of the ion for which they are selective.

(*iii*) Their lower limit of measurement is 10^{-5} mol dm^{-3}.

(*iv*) The lower limit of measurement is independent of pH.

3.2.6. Stability and Lifetime

Solid-state membrane electrodes are among the most stable of all ion selective electrodes. In solutions which are well stirred and with adequate temperature control, the fluoride ise will drift less than 0.1 mV h^{-1}. The slope of such electrodes is also constant over several months of regular use.

Π Two main problems affect the stability of solid-state membrane electrodes. Can you suggest what they are?

We have encountered the first of these problems already.

(*i*) If the electrode is used regularly with solutions that contain an interfering ion, a surface film of precipitate will build up on the surface. This will seriously increase response time and stability.

(*ii*) The second problem also involves a change in the electrode surface. After regular use, particularly with dilute solutions, some of the membrane will dissolve. Unfortunately, it will not dissolve uniformly and the surface will become pitted as illustrated in Fig. 3.2e. Pitting adversely affects both the reproducibility and response time of the electrode. Small volumes of solution will be trapped in the cavities and transferred from one sample to the next.

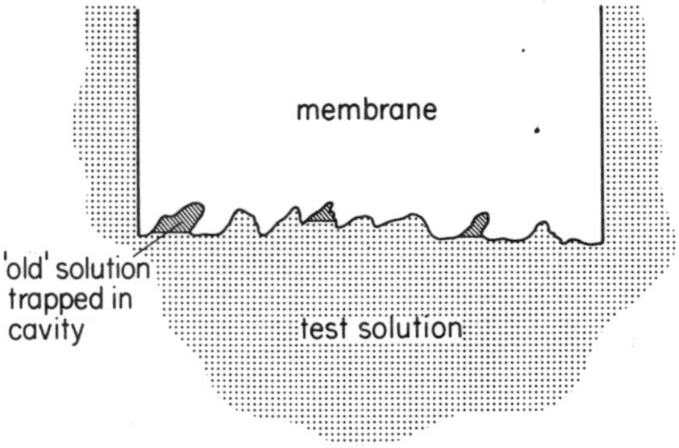

Fig. 3.2e. *Trapping of solution in cavities giving rise to sample carry over when electrode is pitted*

To overcome both of these problems it is advisable to polish the surface of the membrane with a fine grade powder, such as 0.25 μm bauxite polishing powder or similar. This allows a fresh, smooth surface to be presented to the test solution. The frequency of polishing will depend upon usage of the electrode and an experienced user will spot the early signs of drift and response times.

If used routinely, solid state membrane electrodes will give excellent results for two to three years. If used intermittently, eg for student experiments in a teaching laboratory, they can last much longer. On the other hand, if they are used in solutions containing substances that react with the membrane, the lifetime is very much reduced. For example, if the I^- ise with an AgI membrane is stored in a 1.0 mol dm^{-3} solution of KI, it will cease to behave as an ise after about 12 hours.

SAQ 3.2g

Describe two types of connection at the inner surface of a solid state ise. How do they function as inner reference solutions?

SAQ 3.2h

Lead chromate is less soluble than lead sulphide, yet the latter is used with a silver salt to prepare a disc for a Pb^+ ise. Why is this the case?

SAQ 3.2i

Suggest how you could construct a solid state ise which would respond to CN^- ions in solution.

Summary

Solid-state membrane electrodes are made from either a single ionic crystal or a compressed disc. In some cases some inert material, such as poly(ethene) is added for strength. They are among the most stable and reproducible of all ion selective electrodes. They obey the Nernst equation with theoretical slopes being obtained on calibration. In routine use, they will last for at least two years and, if well cared for, very much longer.

They respond to the ion to be determined by establishing a boundary potential due to a solubility equilibrium at the surface. The difference between the inner and outer boundary potentials causes a diffusion potential through the membrane. The selectivity of the membranes works on the principle that only ions of the same size, shape and charge as the ion under investigation can occupy lattice sites and establish an equilibrium. The combination of these three factors is generally unique for a given ion and it makes the electrodes highly selective.

Interferences are generally caused by surface reactions contaminating the surface of the electrode and releasing free ions into solution. Surface reaction occurs if a compound less soluble than the membrane material is formed.

Objectives

You should now be able to:

- sketch the three types of solid state membrane electrodes;
- state the reasons why heterogeneous membranes are used;
- describe how internal solution and all solid state configurations act as internal references;
- describe how the outer boundary potential is achieved for single and mixed compound membranes;

- describe how charge transfer across the membrane occurs via a defect mechanism;

- state the factors influencing upper and lower limits of measurement;

- predict the likely sources of interference;

- state typical response times and electrode lifetimes;

- predict the effect of pH on electrode performance.

3.3. ELECTRODES BASED ON ION EXCHANGE AND NEUTRAL CARRIERS

Overview

In the first two sections of this Part of the Unit (3.1 and 3.2), we have seen how glass and solid state membrane electrodes operate. Electrodes of these two types can only cover a relatively narrow range of ionic species in solution. Whilst glass electrodes respond to certain monovalent cations, and solid state electrodes to a few ions with insoluble salts, there are none that respond to a number of important ions such as NO_3^- and Ca^{2+}.

Electrodes based on ion exchange compounds and neutral carriers cover a very much wider range of compounds than glass or solid state, but their performance in terms of selectivity and lifetime is generally not as good. The three electrodes of this type with the best performance and, consequently, the most frequently used are those for Ca^{2+}, K^+ and NO_3^-.

In this section we will apply many of the ideas introduced earlier to show how we can construct these new types of electrodes. We will also examine what species can interfere with the selectivity of such electrodes and what factors affect their performance.

3.3.1. Construction

The construction of electrodes based on ion exchange compounds and neutral carriers is similar to solid-state electrodes with an internal reference solution. The basic principle is still:

Test solution	Membrane	Internal reference solution

The outer body, like solid state electrodes, is generally made from a robust polymer such as ABS. The main difference is in the actual membrane itself. There are two general types, which are illustrated in Fig. 3.3a.

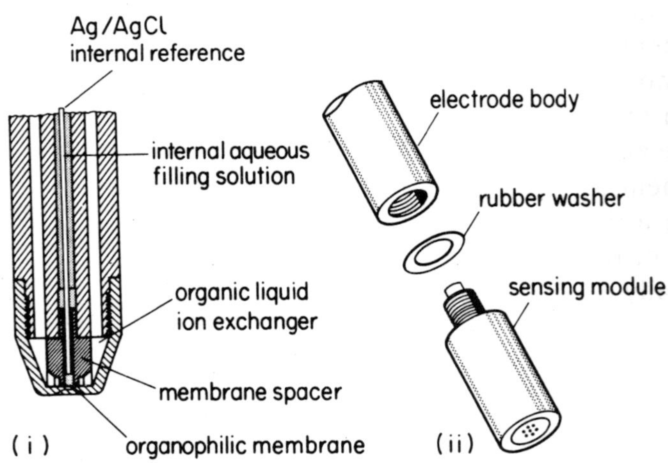

Fig. 3.3a. *(i) Liquid ion exchange membrane electrode, (ii) solid pvc ion exchange membrane electrode built into a sensing module (After Orion)*

The first type (*i*) is termed a liquid membrane. This makes use of an ion exchange compound or neutral carrier dissolved in a water-immiscible organic solvent. This solution is then absorbed into an inert millipore or nucleopore filter. The filter is then held in position at the base of the electrode by a screw cap.

The second type (*ii*) uses a solid membrane, normally made from PVC. The ion exchanger or neutral carrier is then incorporated into this inert matrix. On the surface, solid membranes resemble the heterogeneous solid state membranes described in the previous section. However, the PVC acts as a support medium, rather than as a binder. The amount of PVC in the membrane and the method of manufacture are, therefore, very different. Solid membranes are generally glued to the base of the electrode, forming a non porous seal.

The use of the term solid membrane can sometimes lead to confusion. I will use 'solid membrane' when referring to electrodes using ion exchange or neutral carriers, and the term 'solid-state membrane' for electrodes based on ionic crystals.

A great deal has been written about the relative merits of liquid and solid membranes. The method of operation of the two types is identical and this results in no difference in selectivity. Liquid membranes generally last longer, drift more and respond quicker than their solid counterparts. The older types of liquid membrane were also difficult to assemble and it was this that lead to the development of the solid membrane. However, modern electrodes come with a throw away module (see Fig. 3.3b), which screws onto the end of the electrode body. This makes the assembly of either type equally easy.

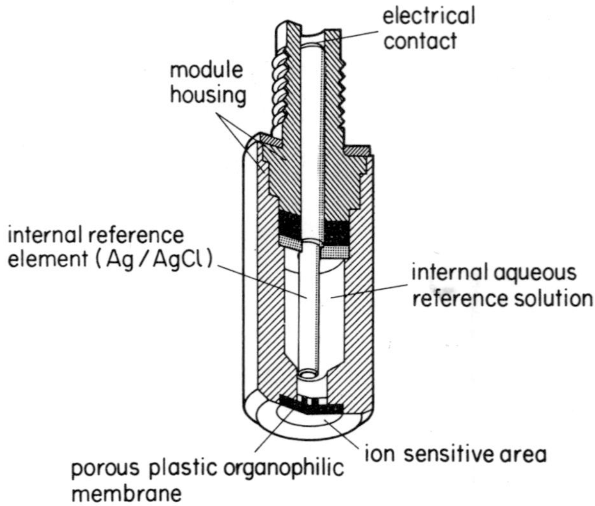

Fig. 3.3b. *Cut-away diagram of the throw-away sensing module for the Orion nitrate ion-selective electrode*

As a general rule, if an electrode is used regularly for routine analysis or continuous monitoring, it is better to opt for the longer life of a liquid membrane. For intermittent use or for potentiometric titrations, the solid membrane gains favour.

SAQ 3.3a

Sketch a liquid membrane ion selective electrode, that makes use of an ion exchange compound.

SAQ 3.3a

3.3.2. Method of Operation

The method of operation of ion exchange and neutral carrier electrodes is virtually identical. We can illustrate this by considering Ca^{2+} and K^+ electrodes.

The Ca^{2+} electrode uses an ion exchange principle. Even for the Ca^{2+} ion alone, a number of different compounds can be used as ion exchangers. The majority are based on organic phosphates such as bis-(di-n-decyl) phosphate:

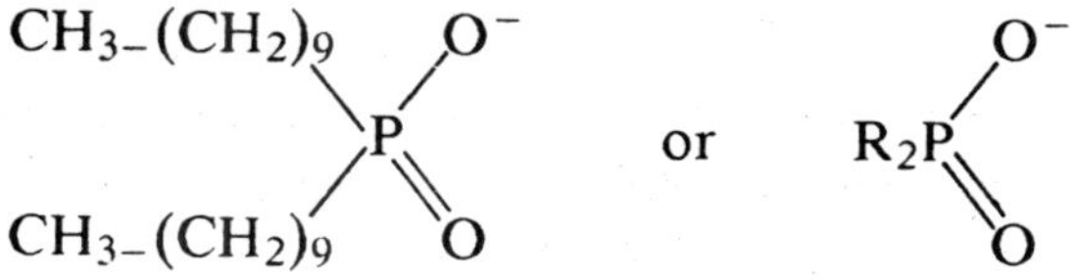

These consists of a long organic chain with a charged phosphate group at the end. At the two surfaces of the membrane, the polar phosphate ends of the molecule will orient themselves towards the water, whilst the non-polar tail will point towards the centre of the membrane as illustrated in Fig. 3.3c. Initially, the phosphate would be in its protonated (acid) form, but once immersed in a solution of Ca^{2+} ions, it would ion-exchange:

$$2\,R_2PO_2H(\text{membrane}) + Ca^{2+}(aq) \rightleftharpoons (R_2PO_2)_2^{2-}Ca^{2+}\ (\text{membrane}) + 2\,H^+(aq)$$

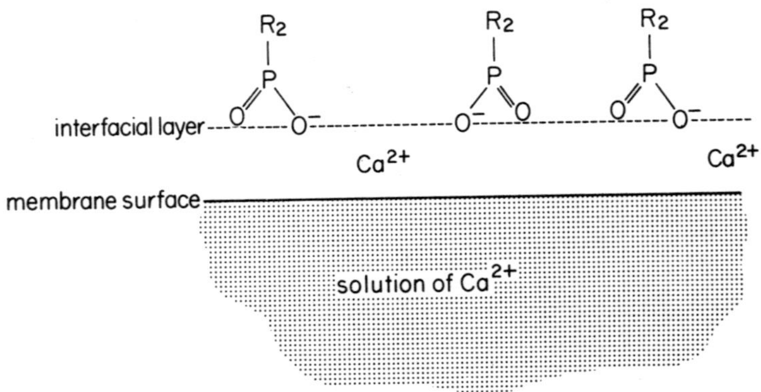

Fig. 3.3c. *Illustration of ionic attraction between calcium ions and ion-exchange molecules at surface of membrane*

The exchange would take place at both surfaces of the membrane, and once conditioned, the electrode would present a surface of Ca^{2+} ions to the internal and test solutions.

Let us now consider the inner surface. There will be Ca^{2+} ions on the surface of the membrane and in the internal solution. A boundary potential will be established due to the differences in activity between the Ca^{2+} ions in the two locations. As the activity of the Ca^{2+} ions in the internal reference solution is constant, the magnitude of the inner boundary potential will also be a constant. This then forms the internal reference, in the same way as with the glass electrode.

The outer surface of the membrane will be in contact with the test solution and the magnitude of the boundary potential here will be a function of the calcium activity in the test solution. Thus, overall, the potential of the electrode will be governed by the activity of Ca^{2+} in the test solution. We can represent the response of the electrode by the Nernst equation, remembering that the Ca^{2+} ion is divalent:

$$E(\text{cell}) = E^* + 0.0296 \log a(Ca^{2+})$$

where E^* incorporates the potentials of the internal and external references as before.

Π How is the electrode selective?

The selectivity of this electrode depends on the high stability of the calcium bis-(di-n-decyl)phosphate complex. The stability of the bis-(di-n-decyl)phosphate, when complexed with other cations, such as sodium, potassium and magnesium, is very much less and fewer surface sites will be occupied by these ions.

Π What effect will pH have on the electrode response?

The initial ion-exchange process, when the electrode was conditioned, involved:

$$\underset{\text{(membrane)}}{2\,R_2PO_2H} + Ca^{2+}(aq) \rightleftharpoons \underset{\text{(membrane)}}{(R_2PO_2)_2Ca} + 2\,H^+(aq)$$

Clearly, at low pH values (acid solutions), the equilibrium will move to the left. Fewer surface sites will be occupied by the Ca^{2+} ions and the electrode's response will fall. Therefore, it is advisable to work in the pH range 5.5 to 11. Above pH 11, the high OH^- concentration, will result in precipitation of $Ca(OH)_2$.

Π What effect will the organic solvent in the membrane have on the electrode's performance?

Superficially, it would appear that the organic solvent used to dissolve the ion exchanger, would have no effect on the performance of the electrode. However, solvent effects do influence the stability constants of complexes formed. We can turn this to our advantage. The Ca^{2+} ise descrribed above, generally makes use of a solvent such as di-n-octylphenylphosphonate. In this solution Ca^{2+} forms a strong complex with bis-(di-n-decyl)phosphate, whilst Mg^{2+} does not. If we change the solvent to the less polar decanol, both Ca^{2+} and Mg^{2+} form equally strong complexes with bis-(di-n-decyl)phosphate. This allows us to construct an electrode that responds equally to both ions according to:

$$E(\text{cell}) = E^* + 0.0296 \log [a(\text{Ca}^{2+}) + a(\text{Mg}^{2+})]$$

Such an electrode is ideal for measuring water hardness.

The NO_3^- ise makes use of a similar principle, ie we select a large charged organic grouping coupled with the ion in question. This achieves the objectives of water insolubility and the free ion which is able to exchange at the surface. Thus, for a given ion, there will often be a number of different ion-exchangers which can be used. In the nitrate case, the ion exchanger will utilise a large positively charged organic grouping with an NO_3^- anion, represented as: R^+ NO_3^-

A common grouping of ion exchangers for nitrate are based on quaternary ammonium salts, such as (tri-dodecyl)hexadecyl-ammonium nitrate.

Neutral carriers work in a similar way to the ion exchangers described above. In designing these electrodes, we need to select a neutral compound that will complex with the ion under investigation. Let us consider the K^+ ise which utilises this principle. The antibiotic valinomycin acts as the neutral carrier and at the junction between the membrane and the solution, an equilibrium will be formed according to:

$$K^+(aq) + V(membrane) \rightleftharpoons K^+V(membrane\ surface)$$

where 'V' represents the valinomycin. The half cell can be represented by:

Test solution	Interfacial layer	Membrane	Interfacial layer	Internal solution
K^+	K^+V	V	K^+V	K^+

The boundary potential is set up between the K^+ ions in the solution and the interfacial layer on the surface of the membrane. This is identical to the ion exchange compounds discussed earlier and the electrode responds according to the Nernst equation:

$$E(\text{cell}) = E^* + 0.0591 \log a(K^+)$$

The selectivity of the electrode relies on the fact that valinomycin forms a very stable complex with K^+ ions, but not with any others. The electrode can be constructed using a liquid membrane, where the valinomycin is dissolved in diphenylether, or a solid membrane where a solution of valinomycin in dioctyldipate is incorporated into a PVC matrix.

The possibilities of designing ion selective electrodes using either liquid ion exchange or neutral carriers are endless. In this section, we have only examined some very popular commercial electrodes. However, it is very easy to construct an ise of this type in the laboratory and experiment with different active materials (the ion exchanger or neutral carrier) and solvents. Fig. 3.3d shows a simple (and cheap) electrode which I have used to try out some different combinations.

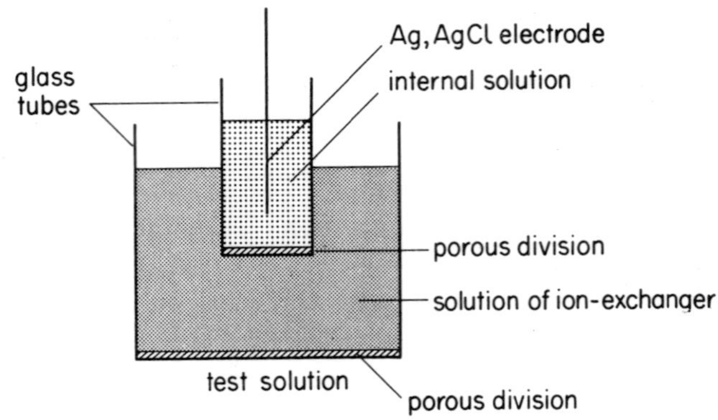

Fig. 3.3d. *Simple electrode using a solution of an ion exchange compound*

SAQ 3.3b

List a few points that are important when selecting materials for ion exchange and neutral carrier ion selective electrodes.

SAQ 3.3c

Which of the following compounds could be used as an active material for a Cl^- ise?

(*i*) teta-octylammonium nitrate;

(*ii*) dioctadecylphosphinous chloride;

(*iii*) chlorovaleric acid;

(*iv*) dimethyl-dioctadecylammonium chloride.

3.3.3. Selectivity

Electrodes based on the compounds mentioned in 3.3.2 are generally less selective than other types of ion selective electrode.

Π Can you suggest how other ions would interfere with the determination of the ion under test?

The selectivity of the membrane reflects the relative stabilities of the complexes the active material makes with interfering ions. Let us consider an ise responding to positive ions – a similar argument can be used for negative ions. If we represent the active material by N, the ion under test by X^+ and the interfering ion by I^+, then we need to examine the equilibrium:

$$\underset{\text{(membrane surface)}}{X^+N} + I^+(aq) \rightleftharpoons \underset{\text{(membrane surface)}}{I^+N} + X^+(aq)$$

If X^+N is much more stable than I^+N, then the electrode will be much more selective towards X^+. If, on the other hand, I^+N is much more stable than X^+N, then I^+ will interfere very strongly with the estimation of X^+. Between these extreme cases, a range of possible interferences can occur. If more than one interfering species is present, then the picture becomes more complicated.

If we examine the stability constants for the equilibria represented generally by:

$$I^+ + N \rightleftharpoons I^+N$$

we can get some idea of the possible level of interference. Remember, it is important that we examine this equilibrium in the solvent used in the membrane. We have already seen with the Ca^{2+} electrode, that changing the solvent can affect the selectivity.

Another, not so common, form of interference can occur if a large organic molecule with a similar charge to the ion under investigation is present in the test solution. Often this is more soluble in the membrane solvent, than in the test solution. It then partitions itself between the membrane and the test solution. In the membrane we now have two species of similar charge. The organic ion will repel the ion under test, reducing its activity in the membrane and affecting the boundary potential. This in turn leads to erroneous values of E(cell).

An example of this type of interference occurs with the K^+ electrode based on valinomycin. The cation of quaternary ammonium salts with large (more than 5 carbons) organic groups, such as hexadecyltrimethylammonium bromide can interfere. This ion has some surfactant properties and diffuses into the interfacial layer of the membrane. As it has a similar charge to the K^+ ion, it effectively repels it from the membrane surface. The activity of the K^+ ion in the membrane is reduced, and that in the test solution is increased.

The level of interference to an ise is generally represented by the selectivity coefficient. With solid-state electrodes, these selectivity coefficients are constant over a range of concentrations and can be measured with some certainty, particularly, if only one interfering species is present. Unfortunately, it is not so easy to obtain values for the selectivity coefficient of electrodes based on ion exchange and neutral carriers. The selectivity coefficients of ion exchange electrodes, in particular, vary independently with the concentrations of the ion under investigation and the interfering ion. Any quoted values of selectivity coefficient should, therefore, be taken as a rough guide only. Selectivity coefficients for electrodes based on neutral carriers, which are slightly more stable, can be relied on slightly more.

As we have stated on a few occasions, the only sure way of deciding the suitability of a electrode for a given application is to calibrate it with test solutions having a similar level of interference.

SAQ 3.3d

Which of the following ions is likely to interfere with the operation of a potassium ise based on valinomycin?

Na^+	$k_{K,Na}$	$= 10^{-4}$
NH_4^+	k_{K,NH_4}	$= 10^{-2}$
Cs^+	$k_{K,Cs}$	$= 10^{-1}$
Rb^+	$k_{K,Rb}$	$= 5$
Li^+	$k_{K,Li}$	$= 10^{-4}$
Ca^{2+}	$k_{K,Ca}$	$= 10^{-5}$
Mg^{2+}	$k_{K,Mg}$	$= 10^{-5}$

SAQ 3.3e

Write a chemical equation to show how a caesium ion, Cs^+, can interfere with a potassium ion selective electrode using the neutral carrier valinomycin (V), which responds to a potassium ion, K^+.

3.3.4. Response Times and Limits of Detection

The response times of these electrodes are much longer than those of the solid state and glass electrodes. Under ideal conditions, electrodes take two or three seconds to achieve 95% response to ten-fold changes in concentration. Different experimenters have quoted response times of between 15 s and 30 min for larger volumes (about 50 cm^3) of sample contained in a beaker. Generally, about 30 s or so is considered a normal response. In the presence of interfering species, or electrodes that have a history of being used with contaminated solutions, the response times are very much greater.

Π What factors influence the upper and lower limit of measurement?

The upper limit of measurement of the majority of liquid membrane electrodes of this type is about 0.1 mol dm^{-3}. The reason is that above this concentration the membrane becomes saturated with the ion under test. With concentrations below 0.1 mol dm^{-3} the ion under test complexes with the active molecules in the interfacial layer. Above this concentration, the ion diffuses into the membrane and complexes with active molecules below the interfacial layer. In successive measurements, these ions deep in the membrane diffuse slowly outwards causing the electrode to drift. Thus, whilst the electrode does respond to solutions more concentrated than 0.1 mol dm^{-3}, it takes a long time to respond and succeeding measurements are inaccurate due to electrode drift. We, therefore, fix the upper limit of measurement at this concentration.

This diffusion of ions into the bulk of the solution, also applies if the membrane has been in contact with a high concentration of an interfering ion. The ion will diffuse out into more dilute solutions creating drift in succeeding measurements.

Ion exhange and neutral carrier electrodes using solid membranes have slightly higher response limits and are less affected by carry-over of either the ion under investigation or an interfering ion. The reason for this is that the active material is much less mobile in the solid membrane and diffusion into the membrane takes very much longer. The electrode will then have been removed from the solution before too much penetration has occurred.

Solid and liquid membranes have similar lower limits of Nernstian response and lower limits of measurement. The limit depends upon the solubility of the active material and its solvent in the test solution. The lower limits of Nernstian response fall in the region 10^{-4} to 10^{-5} mol dm^{-3}. The lower limit of measurement is roughly an order of magnitude lower. Fig. 3.3e summarises the approximate response range of some commercial electrodes.

Ion	Membrane	Response Limits /mol dm^{-3} Upper	Lower	pH Range
Ca^{2+}	Liquid(I)	1	10^{-6}	6–10
Ca^{2+}	Solid(I)	1	10^{-6}	5–9
Ca^{2+}	Solid(N)	1	10^{-6}	4–10
K^+	Liquid(N)	1	10^{-6}	3–10
K^+	Solid(N)	1	10^{-6}	3–10
NH_4	Liquid(N)	10^{-1}	10^{-6}	5–8
NO_3^-	Liquid(I)	1	10^{-6}	4–11
NO_3^-	Solid(I)	10^{-1}	10^{-6}	5–11
Cl^-	Liquid(I)	1	10^{-5}	3–10

Fig. 3.3e. *Response and pH ranges for some commercial ion exchange and neutral carrier ion selective electrodes. (I) refers to an ion exchange membrane and (N) refers to a neutral carrier.*

The pH range over which these electrodes operate varies from membrane to membrane. In general, extremes of pH are avoided. For cation electrodes, H^+ can act as an interfering ion, complexing with the active material. Thus, alkaline pH values are often used, but care must be taken that too high an OH^- ion concentration does not lead to precipitation. For anion electrodes, the OH^- ion can act as an interfering species.

SAQ 3.3f

Which of the following statements are true when applied to ion exchange and neutral carrier electrodes?

(*i*) The electrodes have very short response times.

(*ii*) The lower limit of measurement is comparable with that of solid-state membrane ion selective electrodes.

(*iii*) After coming into contact with a concentrated solution of an interfering ion, the electrode is prone to drift.

(*iv*) The presence of interfering ions increases the response time.

3.3.5. Stability and Lifetime

A great deal has been written on the stability of ion selective electrodes based on ion exchange and neutral carriers. It is difficult to quote exact values as they depend upon the concentrations of the solutions and level of interference. When working with pure solutions, electrode drift can be of the order of 1 mV per hour. The presence of interfering ions will increase this drift as discussed in the previous section. In any event, it should be recognised that this type of ise will drift more than either glass or solid state electrodes.

Π A number of problems effect the stability and lifetime of ion exchange and neutral carrier electrodes. Can you suggest what they are?

We have encountered some of these problems already; they include the following:

(*i*) If the electrode is used with concentrated solutions of the ion under test, the membrane can become saturated. It takes a long time for the excess ion to diffuse out and in some cases saturation can render the electrode useless.

(*ii*) The second problem also involves a change in the electrode surface. After regular use, with a strongly interfering ion, the entire interfacial layer can lose all of the ion for which the electrode was originally selective. The electrode then becomes an ion selective electrode for the interfering ion.

(*iii*) Interfering ions can, not only poison the surface of the membrane, but also diffuse into the bulk of the membrane. This can cause extensive drift until it all diffuses out. As with the membrane becoming saturated with its own ion, presence of an interfering ion in the bulk of the membrane can render it useless.

(*iv*) Two problems that occur irrespective of the concentrations of the test and interfering ion revolve around solubility. If we have an aqueous test solution, then the active material and its solvent will have a limited solubility in water. If a number of measurements are performed with the electrode, the contents of the membrane will slowly leak away. Conversely, the membrane solvent will be able to dissolve a small amount of water, allowing some aqueous ions to be present in the membrane. This will reduce the specificity of the membrane with time.

Whilst all these factors affect both the stability and lifetime of the electrode, the first three tend to affect stability. Under normal usage, it is the fourth factor that tends to limit the lifetime to 2 months. Leakage from solid membranes is less than from liquid membranes, and the former last slightly longer.

If the electrodes are used with non-aqueous solvents, then the solubility of the membrane solvent in the test solution solvent may be much greater and vice versa. This will reduce electrode life, raise the lower limit of measurement and may even prohibit the use of this type of electrode in such systems.

Summary

Ion exchange and neutral carrier electrodes can have either a solid or liquid membrane. The solid membrane is now the more common and has some advantages, notably longer life, easier preparation and less drift. The key to electrode operation is the selectivity of the active material in the membrane for the ion under investigation. If a suitable active material can be found, then an ise may be constructed for any ion. As with other ion selective electrodes, the response is due to a boundary potential set up between the ions in the test solution and those complexed on the surface of the membrane.

Electrodes of this type suffer from interference due to the formation of other complexes by the active material. The selectivity depends upon the relative stability of the complexes formed between the active material and the test or interfering ions. Very high levels of interference can cause diffusion into the bulk of the membrane and cause excessive drift or even failure of the electrode.

The response times, lifetimes and performance of these electrodes are not as good as glass or solid state membrane electrodes. They are, therefore, rarely used if an equivalent solid state electrode is available. However, their one advantage is the wide range of ions that can be covered using them.

Objectives

You should now be able to:

- sketch a typical liquid ion exchange membrane ise;
- state that liquid ion exchange and neutral carriers can use liquid or solid membranes;
- describe the method of operation of liquid ion exchange and neutral carrier electrodes and select suitable materials;
- describe how interfering ions affect the performance of liquid ion exchange and neutral carrier electrodes;
- state the factors influencing response times stability and lifetimes of liquid ion exchange and neutral carrier electrodes.

3.4. GAS SENSING PROBES

Overview

Gas sensing probes are based on the electrodes already discussed and use two membranes rather than one. They are the most selective of all ion selective electrodes, though their response times are somewhat long. They are used primarily for the estimation of dissolved CO_2 or NH_3 in solution, though probes are also available for SO_2, H_2S and nitrogen oxides such as NO and NO_2.

3.4.1. Construction and Theory

The ammonia gas sensing probe is represented schematically in Fig. 3.4a. It consists of an internal cell with a Ag,AgCl reference and pH glass electrodes immersed in a solution of NH_4Cl. This internal cell is separated from the test solution by a hydrophobic gas-permeable membrane. It is this gas permeable membrane that allows ammonia gas to diffuse through selectively. It does not allow the passage of any ions in the test solution to the internal NH_4Cl solution.

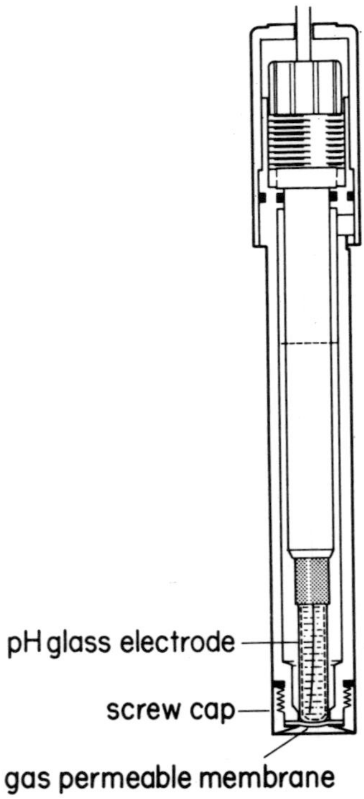

Fig. 3.4a. *Typical ammonia probe (after EIL)*

Ammonia gas dissolved in the test solution will diffuse through the membrane until the partial pressure in the internal solution equals that in the test solution. To avoid the necessity for large amounts of NH_3 to diffuse through the membrane, the volume of the internal solution is kept to a minimum. When the activity of dissolved ammonia has reached equilibrium in the internal solution, it will establish a further equilibrium with the ammonium ions:

$$NH_3(aq) + H^+(aq) \rightleftharpoons NH_4^+(aq)$$

for which the equilibrium constant is:

$$K = a(NH_4^+)/a(NH_3)\ a(H^+)$$

which on rearrangement gives:

$$a(H^+) = a(NH_4^+)/K\ a(NH_3)$$

The internal cell consists of a glass and a Ag,AgCl reference electrode and, consequently, responds to pH changes. At 298 K the response to $a(H^+)$ is:

$$E(\text{cell}) = \text{constant} + 0.0591 \log a(H^+)$$

Substituting for $a(H^+)$ gives:

$$E(\text{cell}) = \text{constant} + 0.0591 \log [a(NH_4^+)/K\ a(NH_3)]$$

The activity of the NH_4^+ ions in the original NH_4Cl solution will be high relative to the activity of NH_4^+ produced by the NH_3 diffusing through the membrane. We may, therefore, consider $a(NH_4^+)$ a constant, which is a function of the original internal solution. As K is also a constant, these two terms can be incorporated into the existing constant in the Nernst equation to give E^*, and the Nernst equation can be written:

$$E(\text{cell}) = E^* + 0.0591 \log [1/a(NH_3)]$$

or

$$E(\text{cell}) = E^* - 0.0591 \log a(NH_3)$$

The gas sensing probe, therefore, responds to the activity of the dissolved ammonia in the external solution.

Other gas sensing probes work in the same way, the only difference being in the choice of internal electrolyte and ion selective electrode. Fig. 3.4b summarises these variables for some commercially available probes.

Test ion	Internal electrolyte	Internal ise	Membrane
NH_3	NH_4Cl	pH	ptfe
CO_2	$NaHCO_3$	pH	ptfe
NO_x	$NaNO_2$	pH	ptfe
SO_2	$K_2S_2O_5$	pH	silicone rubber
H_2S	citrate buffer	S^{2-}	silicone rubber

Fig. 3.4b. *Characteristics of commercial gas sensing probes*

SAQ 3.4a

Sketch a CO_2 gas sensing probe. Write down the equation representing the equilibrium in the internal electrolyte.

SAQ 3.4a

3.4.2. Selectivity and Response Range

The excellent selectivity of gas sensing probes revolves around the use of two membranes. The outer gas-permeable membrane excludes all ions and other non-volatile material. Thus, only dissolved gases can permeate into the inner cell. The gas under test then equilibrates with the inner solution. The membrane of the inner ise is then selective to a product or reactant of the equilibrium established in the inner electrolyte. The only possible source of interference occurs when more than one gas can interact with the inner electrolyte. This is very rare but if it occurs, the interfering gas must be removed from the test solution.

The upper limit of measurement for the NH_3 and CO_2 probes is about 1 mol dm^{-3} whilst for the others it is somewhat less. For the ammonia probe, the upper limit of Nernstian behaviour occurs when the activity of NH_4^+ formed by the diffused NH_3 becomes significant relative to the originally present, ie when $a(NH_4^+)$ ceases to be constant. The upper limit of measurement is, therefore, controlled by the strength of the original NH_4Cl solution. A similar argument can be used for other gas sensing probes.

Π If the concentration of the internal electrolyte affects the upper limit of response, why don't we use a saturated solution?

The more concentrated the internal solution, the greater the osmotic pressure across the membrane. If the pressure is very high, water vapour will diffuse across the membrane, diluting the internal electrolyte. This process of osmosis will be slow and whilst it is taking place, the activity of all species in the internal electrolyte will be changing and the potential of the probe will drift continuously.

The lower limit of measurement of the probes are not always a function of the electrode. It is very difficult to prepare water containing less than 10^{-6} mol dm^{-3} of NH_3. This effectively determines our lower limit of measurement. Similarly, it is difficult to exclude atmospheric CO_2 and obtain concentrations below 10^{-5} mol dm^{-3}. Some have suggested that the lower limit of measurement of both the CO_2 and SO_2 probes are controlled by the generation of the gas by hydrolysis of the internal electrolyte.

Π What pH should be used for the test solution for the NH_3, CO_2 and SO_2 probes?

For the ammonia probe, very alkaline pH values (greater than 12) are needed. This is to prevent the formation of ammonium ions in the test solution according to:

$$NH_3 + H^+ \rightleftarrows NH_4^+$$

For the CO_2 and SO_2 probes, the converse is true and acid pH values (less than 3 and 1 respectively) are needed. This suppresses the hydrolysis of the gases according to:

$$CO_2 + H_2O \rightleftarrows HCO_3^- + H^+$$

SAQ 3.4b

Which of the following (there may be more than one) would cause interference if it was present in the test solution when a gas sensing probe was used to measure the activity of dissolved CO_2?

(*i*) NaCl;

(*ii*) CH_3COOH;

(*iii*) SO_2;

(*iv*) NaOH.

SAQ 3.4c

Why might an ammonia gas sensing probe drift for a long period when in use?

3.4.3. Stability and Lifetime

We have already seen that if the concentration of the inner electrolyte and outer test solutions are substantially different, osmotic pressure will cause transfer of water vapour. This will then contribute to drift and it is advisable to match the total ionic strength of the two solutions. This can be done by diluting the test solution or increasing the concentration of the internal electrolyte.

Another cause of drift is if the temperatures of the inner and test solutions are different. The partial pressure of water vapour is temperature dependent and if the test solution is at a different temperature to the internal electrolyte, drift will occur until temperature equilibrium is reached. Apart from these two causes, gas sensing probes are relatively stable over long periods.

Gas sensing probes have relatively short lifetimes, though they are easily and cheaply recharged. The common cause of failure is due to the membrane. This can generally be attributed to external agencies. If the electrode is used with solutions containing precipitates, then the possibility of the membrane surface becoming coated and the micro-pores becoming blocked exists. Another common cause of failure is magnetic stirring bar hitting and cracking the membrane.

For the optimum performance, gas sensing probes are serviced regularly. The membranes are changed and the internal electrolyte replaced. Care must be taken at reassembly. When the membrane is screwed back on, it is important to get it as close as to the inner electrodes as possible. This is to reduce the volume of the inner electrolyte, facilitating shorter response times. If the membrane is overtightened, then it can be damaged against the internal electrodes and need replacing, even before it's used.

With luck, the outer membrane generally lasts, on average, about two months. Some experimenters change membrane routinely every month, whilst others have reported several months continuous usage.

SAQ 3.4d

What is the main reason for failure of gas sensing probes?

Summary

Gas sensing probes make use of an internal cell consisting of an ion selective and a reference electrode. A gas-permeable membrane allows the gas under test to pass into the internal electrolyte. The gas then sets up an equilibrium with the internal electrolyte and the inner ise responds to either a product or reactant in this equilibrium.

Gas sensing probes are highly selective and suffer little from any form of interference. Some pH control or sample treatment is occasionally necessary.

The electrodes are relatively stable in use, suffering drift due to osmotic pressure across the membrane. This can be minimised by effective control of ionic strength and temperature. Membranes generally last for a couple of months and are easily replaced.

The most common dissolved gases analysed by gas sensing probes are NH_3, SO_2, CO_2, H_2S and oxides of nitrogen.

Ojectives

You should now be able to:

- sketch a typical gas sensing probe;
- describe how gas sensing probes give a Nernstian response to the activity of the dissolved gas;
- describe how osmotic pressure can cause drift;
- state the reasons for the response range of the probes;
- predict which species can interfere with a determination;
- state the reasons for failure of the probe.

3.5. ENZYME ELECTRODES

Overview

Like gas sensing probes, enzyme electrodes make use of one of the types of ise discussed previously. The enzyme is used to convert the species under test into an ion for which an ise already exists. As enzymes are specific in their reactions, the analytical process based on them should be highly selective. Whilst this is often true, some procedures suffer from interference problems. There are several different types of enzyme electrode, though in this section we will concentrate on one to illustrate the principle.

3.5.1. Construction and Theory

The urea enzyme ise is typical of electrodes of its class. It was the first enzyme electrode and was developed by Katz and Rechnitz in 1963. There are a few variations and the most common is illustrated in Fig. 3.5a. It consists of an ammonia glass electrode, the surface of which is coated by the enzyme urease. The enzyme is made into a polyacrylamide gel and effectively sets on the surface. Alternatives to the set gel involve making an aqueous suspension of the enzyme and holding it in place around the electrode by a dialysing membrane. Some experimental electrodes make use of the enzyme in a paste which is held in place over the membrane by a piece of nylon stocking.

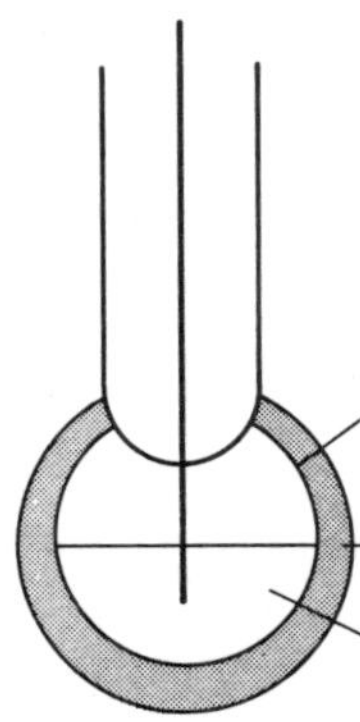

Fig. 3.5a. *Diagram of urea enzyme ion-selective electrode*

The urease is specific in catalysing the hydrolysis of urea:

$$CO(NH_2)_2 + H_2O \xrightarrow{\text{urease}} CO_3^{2-} + 2NH_4^+$$

The NH_4^+ is a quantitative product of the reaction and is a function of the activity of the urea. The ammonium ion diffuses through the gel or solution to the surface of the inner ise. The ammonia-glass electrode responds to the ammonium ion generated and, indirectly, measures the activity of the urea.

The hydrolysis of the urea only occurs at the enzyme surface and very little is used up. The electrode can, therefore, be used repeatedly for analysing a number of solutions. The response of the complete electrode obeys the Nernst equation and at 298 K we can write it as:

$$E(\text{cell}) = E^* + 0.0296 \log a(\text{urea})$$

Enzyme electrodes are very useful when examining biological systems. Some common examples and the associated reactions are listed:

(*i*) Glucose

$$C_6H_{12}O_6 + I_2 + 3\,NaOH \xrightarrow[\text{oxidase}]{\text{glucose}} C_5H_{11}O_5CO_2Na + 2\,NaI + 2\,H_2O$$

The enzyme gel also contains I_2 and we use an I^- ise to monitor the glucose. The presence of an alkaline solution is essential.

(*ii*) Penicillin

$$\text{penicillin} \xrightarrow{\text{penicillinase}} \text{products} + H^+$$

We then use a pH glass electrode to monitor the penicillin.

(*iii*) L-amino acids

$$2\,RCH(NH_3)^+COO^- + O_2 \xrightarrow{\text{L-AAO}} 2\,RCOCOO^- + 2\,NH_4^+$$

where L-AAO represents L-amino acid oxidase. The internal electrode in this case would be the ammonia-glass electrode.

(*iv*) Some workers have taken this a step further and use a mixture or more than one enzyme. One early example was an enzyme electrode to measure the nitrate ion, NO_3^-. The two enzymes were nitrate reductase and nitrite reductase:

$$NO_3^- \xrightarrow[\text{reductase}]{\text{nitrate}} NO_2^-$$

The nitrite ion formed is then reduced:

$$NO_2^- \xrightarrow[\text{reductase}]{\text{nitrite}} NH_4^+ + 2\,H_2O$$

The ammonium ion is detected by an ammonia glass electrode as before.

There is no reason why we should stop at two enzymes, though the response time will increase as the number grows.

SAQ 3.5a

Suggest an internal ise for the following enzyme reaction:

$$\text{L-glutamine} + H_2O \xrightarrow{\text{glutaminase}} \text{L-glutamate} + NH_3$$

3.5.2. Selectivity and Response Range

Π Can you suggest any interference problems when the urea enzyme ise is used with blood samples?

Enzymes catalyse specific reactions and this part of the electrode's response is very selective. Thus, it is only urea that is likely to be converted into NH_4^+.

The major interference problem is due to the presence of Na^+ and K^+ ions in the blood. These ions can diffuse through the enzyme layer, in the same way as the generated NH_4^+ ions. Ammonia glass electrodes have a selectivity order of:

$$H^+ > Ag^+ > K^+ = NH_4^+ > Na^+ > Li^+ > \ldots$$

Clearly K^+ ions, in particular, will create a response, far in excess of that obtained for the generated NH_4^+ ions. One solution to the problem is to remove both the Na^+ and K^+ ions from the blood before measurement. This pretreatment is time consuming and an ingenious solution to this problem was devised by Gilbault. He set up a cell as shown in Fig. 3.5b. He used an uncoated ammonia glass probe as the reference electrode. The difference between the coated and uncoated electrodes was then due to the NH_4^+ generated by the urea.

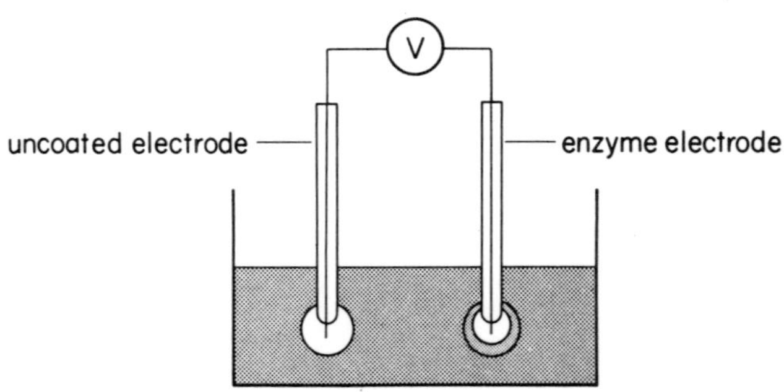

Fig. 3.5b. *Cell using an uncoated ammonia glass electrode as the reference and an enzyme-coated ammonia glass electrode as the indicator electrode*

Other enzyme electrodes suffer from interference in the same way, with the interfering ions diffusing through the enzyme layer from the bulk solution. The two common methods of reducing the effects are to remove the interfering species or to calibrate the electrode using standards with similar levels of interference.

One important point that is often forgotten by those using enzyme electrodes is that enzyme activity can be seriously reduced by traces of silver and mercury.

Π As the common reference electrodes contain either silver or mercury metals how can we minimise contamination of the solution?

I hope you remembered that a double junction reference solution, using saturated KCl or similar in the bridge will virtually eliminate contamination (2.7).

The response ranges of enzyme ion selective electrodes do not follow rigid patterns. Generally, the lower limit of measurement is determined by the response time of the electrode. Once it gets too slow, the limit has been reached. What is considered too slow, depends upon the analyst and the speed of alternative methods. The urea and glucose enzyme electrodes can be used down to about 10^{-4} mol dm^{-3}, whilst the L-amino acid enzyme electrode can be used to about 10^{-5} mol dm^{-3}.

Upper response limits vary from the urea electrode at about 10^{-1} mol dm^{-3} to the glucose electrode at about 10^{-3} mol dm^{-3}. This often leaves us with a relatively small working range, particularly for glucose.

SAQ 3.5b

The glucose electrode (described in 3.5.1.) uses an I^- ise. Which of the following ions would interfere with the determination?

Cl^-;

Na^+;

S^{2-};

Ca^{2+}.

SAQ 3.5b

3.5.3. Response Times and Electrode Life

The response of enzyme ion selective electrodes is much slower than other types of ise. The reason for this is twofold. Firstly, the enzyme catalysed reaction is slow in comparison to the establishment of a boundary potential. Secondly, the time taken for the ion produced to diffuse through the enzyme layer to the inner ise is also significant. Under ideal conditions, a typical enzyme ise should give an almost steady reading after 30 s to 1 minute. The longer the time, the steadier the reading. The time that the electrode is allowed to equilibrate, therefore, has some significance on the accuracy of the final analysis. The shorter the time, the greater the relative error. The longer the time the lower the error. However, doubling the time will not halve the error and it is the law of diminishing returns. Typical relative errors are in the range 2% to 15% depending on time.

The kinetics of enzyme reactions are heavily dependent on temperature and pH. The optimum working temperature is often 37 °C and readings should be taken in a thermostat at this temperature. The electrode is also calibrated at the working temperature and it is important to remember that the slope will not be the same as at room temperature. This is particularly significant if you are using a standard addition procedure. The pH range of enzyme reactions can be very narrow and effective buffering is vital. Unfortunately, the optimum pH for the enzyme reaction may not be the same as the optimum pH for the response of the inner ise. This is one of the major problems in designing good enzyme ion selective electrodes.

There are a lot more variables to consider when using an enzyme ise. Consequently, analytical methods devised must be very detailed, in order that reproducible results are obtained by different analysts. The need to control and reproduce a number of parameters has resulted in many analytical methods for enzyme electrodes being automated. Some of these are described in the final part of the unit.

The useful lifetime of these electrodes is determined by the enzyme layer itself. One coating of enzyme will normally perform a number of analyses before it is eroded away by mechanical action. If the membrane is covered by a thin layer of cellophane, it increases the lifetime, but also increases the response time. Another factor which renders the electrode useless is if the membrane becomes contaminated. The life of an enzyme ise is, therefore, very much determined by use and can range from a few determinations up to a month or so of continuous use.

SAQ 3.5c

List the factors which need to be controlled for accurate analysis with an enzyme ise.

Summary

Enzyme ion selective electrodes make use of a normal ise whose membrane is coated with an enzyme. This enzyme can either be in solution or in the form of a gel and it reacts with the solution to produce an ion or gas which can be detected by the inner ise.

Their response times are long and their lifetimes can be quite short. Further, their conditions of operation need close control for good results. However, they do allow us to obtain direct measurements for a number of biologically significant molecules, which can often be quicker than some spectrophotometric methods needing considerable sample pretreatment.

The main source of interference is from ions diffusing through the enzyme from the bulk of the solution to surface of the inner ise. The selectivity of the electrode is, therefore, that of the inner ise, rather than that of the enzyme layer.

Objectives

Your should now be able to:

- sketch a typical enzyme ise for the estimation of urea and describe its response;
- select a suitable inner ise for an enzyme electrode;
- describe the conditions necessary for accurate readings and predict interfering species;
- state that the response time of enzyme electrodes is long.

3.6. ION-SELECTIVE FIELD EFFECT TRANSISTORS (isfet)

Overview

Over the last decade or so, ion-selective field effect transistors (isfet) have been developed. This section deals very briefly with their construction and their advantages and disadvantages over more conventional types of ion-selective electrode.

3.6.1. Construction

Ion-selective field effect transistors make use of a conventional ion-selective electrode membrane, ie solid state, liquid ion exchange, neutral carrier etc. This thin membrane (approx. 100 μm) replaces the metal gate of a metal oxide semiconductor field effect transistor as illustrated in Fig. 3.6a. A potential is developed across the membrane in the same way as described in the earlier sections. This potential, in turn, controls the current flowing through the transistor. Thus, the current flowing is proportional to the logarithm of the activity of the ion under investigation.

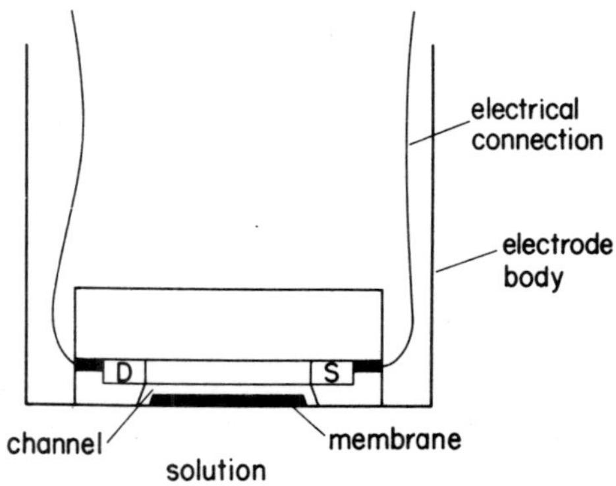

Fig. 3.6a. *Simplified diagram of an isfet. The potential across the membrane governs the current flow between the drain (D) and source (S) through the channel*

Ion selective field effect transistors are generally used with conventional reference electrodes, though recent developments include reference electrodes built into the isfet itself. In fact, this latter type makes use of two isfet's in one body, the second transistor being responsive to a species whose activity is kept constant, eg H^+ ions in a buffer solution.

3.6.2. Selectivity and Lifetime

In principle, the selectivity of an isfet is governed by the membrane and should be no different from conventional electrodes. Early prototypes did not exhibit such large operating ranges as conventional electrodes and calibration graphs had only short linear portions. The behaviour of more recent isfets is very much better, though still not as good as conventional electrodes.

The lifetime of an isfet is generally much less than that of a conventional electrode. Solid state membranes tend to peel off the transistor surface whilst the small volumes of the liquid ion exchange and neutral carrier electrodes result in them becoming contaminated very quickly. It is this small volume which, paradoxically, is the main advantage of ion selective field effect transistors. They can be used with very small volumes of solution and are ideal for biological or biochemical analysis. To increase electrode life, the membrane is now coated in the same way as enzyme electrodes. This gives maximum lifetimes of about 3 weeks to a month.

When using an isfet it is necessary to condition it by immersing in a solution of the appropriate ion for between one and five days. This is necessary to obtain a steady reading and linear calibration. After conditioning, the electrode should be stored above, rather than in, the conditioning solution. This is to prevent leakage of the membrane material and reduction of the operating life.

SAQ 3.6a

> Which of the following factors is the main advantage of ion-selective field effect transistors?
>
> (*i*) more effective membranes;
>
> (*ii*) larger working range;
>
> (*iii*) long lifetime;
>
> (*iv*) small scale.

Summary

Ion-selective field effect transistors are a new development in the area of ion-selective electrodes. Their main advantage is their small size, but short lifetime and non-linear calibration will need to be overcome before they develop into routine analytical tools.

Objectives

You should now be able to:

- sketch a typical ion-selective field effect transistor and describe briefly its mode of operation;
- list the advantages and disadvantages of ion-selective field effect transistors.

4. Applications

Overview

This final Part of the unit deals with some common and not so common uses of ion selective electrodes. They will be set out in the form of case studies. In this way you will be able to see what the requirements and possible pitfalls are. You can then suggest how the analysis could be performed before reading how we actually carry it out.

It would also be very helpful if you could try to carry out some of these experiments yourself.

4.1. CASE STUDY 1. ESTIMATION OF F^- IONS ADDED TO POTABLE WATER

Potable water is the term used in the water industry for drinking water. In some areas of the country, the drinking water has fluoride (F^-) ions added to help prevent tooth decay. The fluoride is added after the water treatment process and is in the region of 1 ppm (ie 1 mg dm^{-3}). The purpose of the analysis is to check that the correct quantity of fluoride has been added. Some of the requirements of the analysis are:

ion to be detected	F^-
concentration	1 ppm
number of samples	15 per day
volume of sample available for test	no constraint
accuracy required	± 5%
sample also contains	Na^+, K^+, Mg^{2+}, Ca^{2+}, Al^{3+} Fe^{2+}, Fe^{3+}, Cl^-, NO_3^-.

Firstly we must select our ise, reference electrode system.

For the F^- ion, we have very little choice. The solid state (LaF_3) membrane electrode is so effective, no commercial manufacturer offers an alternative. As this ise does not suffer from Cl^- interference, a single junction Ag,AgCl reference electrode is acceptable.

We now select a suitable method.

As we have fifteen samples a day to process, standard addition and titration methods would be too time consuming. The use of a calibration graph is by far the most efficient method. We set up one calibration at the beginning of the day, which we can then use for each of the samples.

If you are lucky to have an ion-meter, which plots the calibration graph for you, so much the better. Your calibration standards should span the likely values of your samples. In this case, a calibration range from 0.1 to 10 ppm, would be ideal. As we want to find the total concentration (not activity) of the fluoride added, we will need to add an ionic strength adjuster (isa) to each standard and sample. This method should give results to better than 5% with the F^- electrode.

We also need to examine for possible interferences.

The only ion that interferes with the F^- ise is OH^-. If we buffer the solutions at the electrode's optimum response range between pH 5 and 6, this should pose us no problems. However, whilst the dissolved ions present do not interfere with the electrode itself, some react with the fluoride ions. For example, Al^{3+} forms a complex, $(AlF_6)^{3-}$ with F^- ions. The fluoride ise only measures free F^- ions and not complexed F^-.

Π How do we release the F^- ions from the complex?

We add another complexing agent that forms a stronger complex with the mineral ion. For example, there are a number of complexones such as edta (ethylenediaminetetraacetic acid) which form very strong complexes with Al^{3+} and similar mineral ions. If we use Y^{4-} to represent the edta anion, the reaction between its diprotonated form (H_2Y^{2-}) would be:

$$AlF_6^{3-} + H_2Y^{2-} = AlY^- + 2H^+ + 6F^-$$

The H^+ ions produced would be mopped up by the buffer and the pH of the sample would not change.

We incorporate the buffer and the complexing agent into the isa. It is then given the title total ionic strength adjuster buffer (tisab).

Are there any temperature effects?

The slope of the calibration graph is dependent upon the temperature. Remember that slope $= (2.303RT/nF)$. Thus, all standards and samples must be at the same temperature for accurate analysis. Generally, thermostatting is not necessary if all solutions are allowed to equilibrate to room temperature. Water coming out of a tap is usually cold and it will need a few minutes to reach room temperature before any measurements are taken.

Let us now look at the details of the actual analysis.

(*a*) Prepare standard solutions of F^- ions in the range 0.1 to 1.0 ppm using Analar NaF.

(*b*) Prepare tisab. Orion recommend the following: place approximately 500 cm^3 of distilled water in a 1-litre beaker. Add 57 cm^3 of glacial ethanoic acid, 58 g of sodium chloride and 4 g of cdta (cyclohexylenedinitrilotetraacetic acid) Stir to dissolve and cool to room temperature. Adjust the pH to between 5.0 and 5.5 using 5 mol dm^{-3} NaOH. Pour the mixture into a 1-litre standard flask and make up to the mark.

(*c*) Dilute 25 cm^3 of each standard and sample solution with an equal volume of the tisab and allow each to equilibrate to room temperature.

(*d*) Measure the cell emf of each when the F^- ise/reference electrode assembly is immersed in it.

(*e*) Plot a graph of measured emf against concentration for each standard using semilogarithmic graph paper.

(*f*) Read off the concentration of each sample from the calibration graph.

This method of fluoride analysis is very successful and can be extended to a number of different systems. We can use it for sea and waste waters, to estimate the HF concentration of etching baths and F^- in toothpaste. A standard addition method is used to measure the amount of F^- in petrol and in the estimation of stability constants for fluoride complexes.

SAQ 4.1a

What is the purpose of adding cdta to the isa for the determination of F^-?

SAQ 4.1a

SAQ 4.1b

A student calibrated a fluoride ion-selective electrode with the following results

concentration/ppm	emf/mV
1.00	96.0
5.00	70.5
10.00	59.0
50.00	32.5
100.00	21.5

Draw a calibration graph and estimate the concentration of samples which gave the following emf readings under the same condition: 73.0; 69.5; 42.0; and 63.5 mV.

SAQ 4.1b

4.2. CASE STUDY 2. ESTIMATION OF Ca^{2+} CONCENTRATION IN BEER

Beer is a very complex solution containing a large number of dissolved metal ions and organic compounds in solution. Originally, beer was brewed with local water, notably Pennine water in Yorkshire and Trent water in Burton-on-Trent. Nowadays, tap water is demineralised by an ion exchange process and the minerals added in appropriate quantities to simulate the local water. Even some American beers are brewed with 'Trent' water.

In the analysis of the final brew, the concentration of many of these minerals is checked. The specification for the analysis will include:

ion to be detected	Ca^{2+}
concentration	100 ppm
number of samples	1 per day
volume of sample available for test	no constraint
accuracy required	$\pm$ 5%
sample also contains	a number of monovalent and divalent cations, also calcium complexing agents, including proteins.

Firstly we must select our ise, reference electrode system.

For the Ca^{2+} ion, we have a choice of electrodes. There are a number of solid and liquid membrane ion exchange and neutral carrier electrodes.

Strong interfering species can diffuse into the centre of the liquid membrane and eventually cause drift. The neutral carrier electrode is not selective enough and, in this application, a solid membrane ion exchange ise is the best choice. As with the F^- ise, the Ca^{2+} ise does not suffer from Cl^- interference and a single junction Ag,AgCl reference electrode can be used.

We now select a suitable method.

We have very few samples to analyse and a high level of interference. It is particularly difficult to decomplex the Ca^{2+}. The multiple standard addition method is then chosen and the results plotted using Gran's plot paper (2.6.2).

Do we need any sample pre-treatment?

With the multiple standard addition method, we do not need to worry too much above removing interfering ions. Some sources suggest that the beer should be filtered prior to analysis, though this is not strictly necessary. Any particles present are generally large enough not to clog the membrane.

The only pre-treatment necessary is to buffer the beer at the optimum working range of the ise; between 5 and 6 pH units.

Are there any temperature effects?

As we are performing all measurements on the same solution, if the original solution was at room temperature, there should be little change during the addition processes.

Le us now look at the details of the actual analysis.

(*a*) Prepare a standard solution of Ca^{2+} of concentration about 100 ppm.

(*b*) Pipette 10.0 cm^3 of the beer into a beaker and adjust the pH to between 5.5 and 6.0 using NaOH.

(*c*) Immerse the Ca^{2+} ise/reference electrode into the beer and measure the emf. Add 1.0 cm^3 of the standard Ca^{2+} solution. Stir the solution well and measure the steady emf. Repeat with three further additions of the same standard solution.

(*d*) Plot a graph of emf against volume of added standard solution on Gran's plot paper and estimate the concentration of the beer by extrapolation.

This method of calcium analysis is one of many performed with the Ca^{2+} ise. Calcium is also routinely analysed using atomic absorption spectrophotometry (aas) or ion chromatography. However, if only a few samples are to be measured per week, the ise method will work out cheaper and be as accurate.

SAQ 4.2a

Why is the multiple standard addition method chosen for the analysis of Ca^{2+} in beer?

4.3. CASE STUDY 3. ESTIMATION OF NO_3^- IN PLANT TISSUE

In the first two case studies, we have considered ions that were already in solution prior to analysis. In this example, the ion is contained in plant tissue and will need to be extracted prior to measurement. The analysis needs to measure the total nitrate present as a percentage by weight.

The specification for the analysis will not be as detailed as some, but will include:

ion to be detected	NO_3^-
concentration	10 ppm after extraction
number of samples	groups of 10 to 20
volume of sample available for test	no constraint
accuracy required	as good as possible!
sample also contains	a number of monovalent and divalent cations and anions; in particular Cl^-.

Firstly we must select our ise, reference electrode system.

For the NO_3^- ion, we use an ion exchange type of ise. As it will be used intermittently, the ion exchanger is best contained in a solid membrane. NO_3^- electrodes suffer interference from Cl^- ions. It is, therefore, not advisable to use either a calomel or Ag,AgCl reference electrode. It is better to use a mercury(I)sulphate double junction reference with saturated sodium sulphate as the bridge solution. There are some alternatives to this choice of reference electrode, but each has its own small problem.

We now select a suitable method.

We have a large number of samples to run through in one go. A calibration graph is, therefore, the best method.

How do we get the nitrate out from the plant tissue?

This is the key to the whole analysis. A known weight (about 10 g) of tissue is pulverised and placed into a known volume (about 50 cm^3) of an extracting solution. This is stirred well for some 20 minutes or so to release the nitrate. The extracting solution contains silver sulphate, which prevents chloride in the tissue dissolving as free Cl^-. It is also buffered to pH 3, which is the optimum pH for the nitrate electrode.

The standard solutions for the calibration are also made up in the same extracting solution. Thus, the calibration and measurements are performed in similar solutions. To minimise liquid junction potentials, some workers have suggested that the bridge solution of the double junction reference electrode should be the same extracting solution.

Are there any temperature effects?

As with the two previous examples, it is sufficient to allow all solutions to equilibrate to room temperature.

We will not look at the details of the method this time. It involves preparing a calibration graph using standards prepared in the extracting solution and comparing the samples with it. The results of this method have been compared with a classical Devarda method. Excellent agreement was obtained (correlation coefficient 0.99) with the added advantage that the ise method was quicker.

It is interesting to note that a number of independent workers have compared the NO_3^- ise to classical methods of nitrate analysis and have achieved excellent agreement.

SAQ 4.3a

Why shouldn't a normal Ag,AgCl reference electrode be used with a nitrate ise?

SAQ 4.3a

4.4. CASE STUDY 4. ESTIMATION OF Na^+ OR Cl^- IN BOILER FEED WATER

There are a number of different applications where water is converted into steam in a boiler. In many cases, such as boilers in power stations, the steam is at a temperature of several hundred degrees Celsius. If any electrolyte is present in the water feeding the boiler (the feed water), then galvanic cells can then be set up within the boiler pipes. This will lead to accelerated corrosion and failure of the boiler.

Power stations are often constructed near the coast as the sea water forms a ready source of coolant for the condensers at the end of the turbines. With time, faults can appear in the condensers and some of the sea water can find its way into the feed water. It is imperative that the leak be spotted immediately. This calls for continuous monitoring of the feed water for Na^+ or Cl^- concentration. Whilst the actual concentration of the ion is not of importance, we need a system that can detect as low a value as possible.

The specification for the analysis will not be as detailed as some, but will include:

ion to be detected	Na^+ or Cl^-
concentration	as low as possible
number of samples	continuous monitoring
volume of sample available for test	no constraint
sample also contains	no interfering agents

Firstly we must select our ise, reference electrode system.

We have the option of monitoring either the Na^+ or Cl^- ions. As continuous monitoring is to be used, the membrane will be faced with a fresh solution regularly. Thus, any membrane chosen must be hard-wearing and of limited solubility. The Na^+ ise uses a modified glass membrane which will last for a very long time. The Cl^- ise, on the other hand, uses a solid state membrane of $AgCl/Ag_2S$. This has a limited solubility and will eventually dissolve away. Though, even with continuous monitoring, the electrode should last over a year in service.

Another factor to be considered when selecting the ise, is the lower limit of measurement. Depending on the actual conditions of measurement, the Na^+ ise has a lower limit between 10^{-5} and 10^{-8} mol dm^{-3}. The addition of ammonia gas or certain amines to control the pH allows the lower value to be reached. The chloride ise has a similar lower limit of measurement over a wide pH range.

Both electrodes give stable potential readings over long periods and once a day calibration is usually sufficient.

As you can see, there is very little to choose between either electrode and continuous monitors using either electrode are available. I will select the Na^+ system as it brings out some different features.

Fig. 4.4a shows a schematic diagram for an industrial monitor using a sodium glass electrode. The monitor draws a small amount of water from the boiler feed and treats it with a reagent such as di-isopropylamine to control the pH. To avoid any temperature effects, the flow is then passed through a thermostat, before passing into the measuring cell.

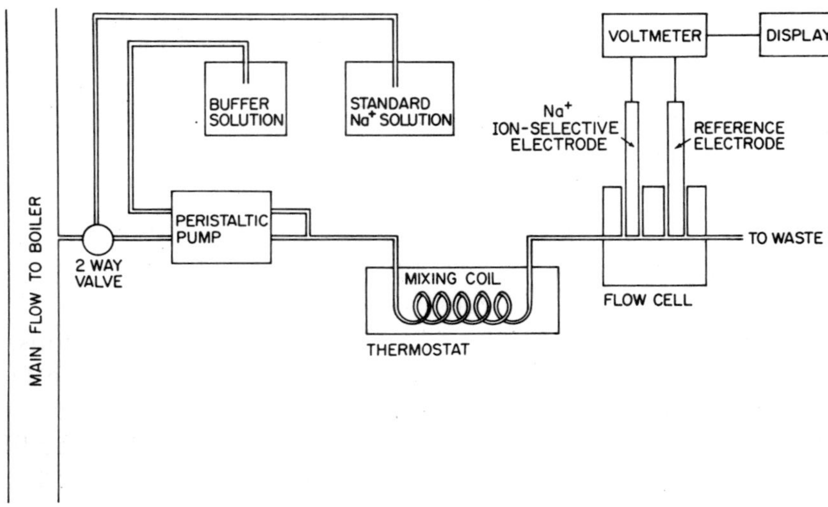

Fig. 4.4a. *Schematic diagram of an industrial monitor using a sodium glass electrode*

Several designs of measuring cell have been tried. Some are very simple, whilst others have greater degrees of sophistication. One type uses a built-in magnetic stirrer, which allows the electrode to respond quicker. However, the added complexity of these designs generally leads to more problems than they solve.

Π A common feature of all these measuring cells is that the reference electrode is placed on the waste side of the cell. Can you suggest a reason for this?

One of the problems with reference electrodes is leakage of the bridge or internal filling solution. This can contaminate the test solution and great care must be taken in selecting reference electrodes when measurements are taken in a beaker. However, in a flow system, any contamination will be washed away and if the reference is on the waste side of the measurement cell, it will not affect the ise.

The potential of the ise/reference system is monitored using a voltmeter in the normal way. In certain applications, a chart recorder can be added in place of, or in parallel with, the meter to give hard copy of the variation in Na^+ concentration over a day. We can extend this continuous monitoring to control a manufacturing process via a feedback loop. For example, in the manufacture of Coca-Cola, measurement of pH can be used to regulate the relative quantities of certain ingredients.

Most continuous monitors of this type make use of an internal standard solution which can be used for automatic calibration. A two-way valve switches from the external flow to the internal standard. This standard is treated in the same way as the sample, it is treated with a reagent to bring it to the optimum conditions for measurement and then brought to an optimum temperature in a thermostat. The frequency of calibration will depend on the ise and the degree of interference in the solution being analysed.

This principle of automatic monitoring can be extended to any system which can be analysed using an ise. It is more expensive than a manual system if only a small number of measurements are to be performed, but is ideal for continuous or frequent monitoring of flow processes.

SAQ 4.4a

Which of the following factors would be advantageous in selecting an ise for continuous monitoring?

(*i*) long lifetime;

(*ii*) low coefficient of thermal stability; ⟶

SAQ 4.4a (cont.)

(*iii*) optimum pH operating range is identical to that of the sample;

(*iv*) smooth membrane surface.

4.5. CASE STUDY 5. ESTIMATION OF L-TYROSINE

The enzyme tyrosine decarboxylase can be used to catalyse the reaction:

$$\text{L-tyrosine} \xrightarrow[\text{decarboxylase}]{\text{tyrosine}} \text{tyramine} + CO_2$$

This enzyme is highly selective and possible interference is unlikely. The main requirements of the analysis are:

species to be detected	L-tyrosine
concentration	10^{-3} mol dm^{-3}
number of samples	5 per day
volume of sample available for test	10 cm^3
accuracy required	$\pm 10\%$
sample also contains	a number of other amino acids

This is clearly a case for an enzyme electrode as no ise would measure the activity of L-tyrosine directly. The enzyme tyrosine decarboxylase would be coated onto the membrane of an ise and held in place with a dialysis membrane.

Π Suggest an ise which could be used for this estimation.

As CO_2 gas is produced in the reaction, a reasonable choice would be a CO_2 gas-sensing probe. The choice of this sensor also solves the problem of selecting a reference electrode as it is built into the probe.

An enzyme electrode of this type has been used by a number of workers and response times of about 5 min have been obtained for the concentration range 10^{-4} to 10^{-2} mol dm^{-3}. The electrode response is linear over this range and the electrode has a near Nernstian slope of 55 mV per decade.

The highly selective nature of the enzyme and the CO_2 probe means that there will be little likelihood of interference.

Π Suggest a suitable method for the analysis of 5 samples per day.

The long response time of the electrode means that using a standard addition procedure for each sample would be very time consuming.

Similarly, the use of Gran's plot methods would also be time consuming and unnecessary considering the selective nature of the system.

We are then left with a calibration graph which could be plotted using two or three standards. This is marginally less time consuming than the standard addition. It also has the advantage that if the slope deviates from 55 mV per decade over several days, it does not affect the analysis.

SAQ 4.5a

> If the number of samples of L-tyrosine per day were reduced to one or two, what steps would you change in the analysis?

Self Assessment Questions and Responses

SAQ 1.1a

Calculate the total ionic strength of a 0.100 mol dm^{-3} solution of sodium sulphate(VI) (note that the formula of sodium sulphate(VI) is Na_2SO_4 and there will be twice as many sodium ions as sulphate(VI) ions).

Response

Sodium sulphate(VI) dissociates

$$Na_2SO_4(s) \rightarrow 2\,Na^+(aq) + SO_4^{2-}(aq)$$

$$\qquad\qquad 0.200 \qquad\qquad 0.100$$

The total ionic strength is calculated as follows:

$$\begin{aligned} I &= 0.5 \sum c_i z_i^2 \\ &= 0.5\,[(0.200 \times 1^2) + (0.100 \times 2^2)] \\ &= 0.5\,[(0.200) + (0.400)] \\ &= 0.300 \text{ mol dm}^{-3} \end{aligned}$$

****

SAQ 1.1b

State the two factors which influence the extent of the ion atmosphere surrounding a given ion.

Response

The two major factors affecting the extent of the ion atmosphere are the charge density of the ion and the total number of ions present (the concentration).

****

SAQ 1.1c

What term is used to reflect the 'effective concentration' of a given ion?

Response

The 'effective concentration' of an ion is quoted as its activity. The activity is related to the concentration by the activity coefficient (γ) so that:

$$a = \gamma c$$

****

SAQ 1.1d

Calculate the activity of the copper ion in a 0.250 mol dm^{-3} solution of copper(II) chloride($CuCl_2$).

Response

The total ionic strength is calculated as follows:

$$I = 0.5 \sum c_i z_i^2$$

$$= 0.5 \, [(0.250 \times 2^2) + (0.500 \times 1^2)]$$

$$= 0.5 \, [(1.000) + (0.500)]$$

$$= 0.750 \text{ mol dm}^{-3}$$

The activity coefficient is calculated:

$$\log \gamma(Cu^{2+}) = -0.5091 \, Z^2 \, (I)^{\frac{1}{2}}$$

$$\log \gamma(Cu^{2+}) = -0.5091 \times 2^2 \times (0.750)^{\frac{1}{2}}$$

$$\gamma(Cu^{2+}) = 0.0172$$

The activity is then calculated:

$$a(Cu^{2+}) = \gamma(Cu^{2+}) \, c(Cu^{2+})$$

$$= 0.0172 \times 0.250$$

$$= 4.31 \times 10^{-3} \text{ mol dm}^{-3}$$

****

SAQ 1.2a

Write down the half-equations corresponding to the oxidation and reduction processes for the following overall reactions:

(*i*) $Fe^{2+}(aq) + Ce^{4+}(aq) \rightarrow Fe^{3+}(aq) + Ce^{3+}(aq)$

(*ii*) $Zn(s) + 2\,Ag^{+}(aq) \rightarrow Zn^{2+}(aq) + 2\,Ag(s)$

(*iii*) $Fe(s) + S(s) \rightarrow FeS(s)$

Response

(*i*) $Fe^{2+} \rightarrow Fe^{3+} + e$ —oxidation

$Ce^{4+} + e \rightarrow Ce^{3+}$ —reduction

Note how the electron is transferred from the Fe^{2+} to the Ce^{4+}.

(*ii*) $Zn \rightarrow Zn^{2+} + 2e$ —oxidation

$Ag^{+} + e \rightarrow Ag$ —reduction

(*iii*) $Fe \rightarrow Fe^{2+} + 2e$ —oxidation

$S + 2e \rightarrow S^{2-}$ —reduction

****

SAQ 1.2b

Draw the galvanic cell represented by the following cell notation.

$Pt \mid Fe^{3+}, Fe^{2+} \parallel Cu^{2+} \mid Cu$

Response

Your sketch should look like:

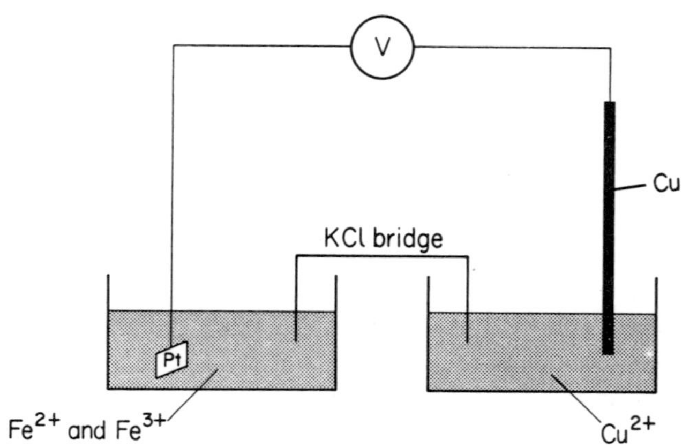

Fig. 1.2d. *Galvanic cell for notation given*

****

SAQ 1.2c

Draw the cell for the determination of the sep of the iron(III)/iron(II) equilibrium.

Response

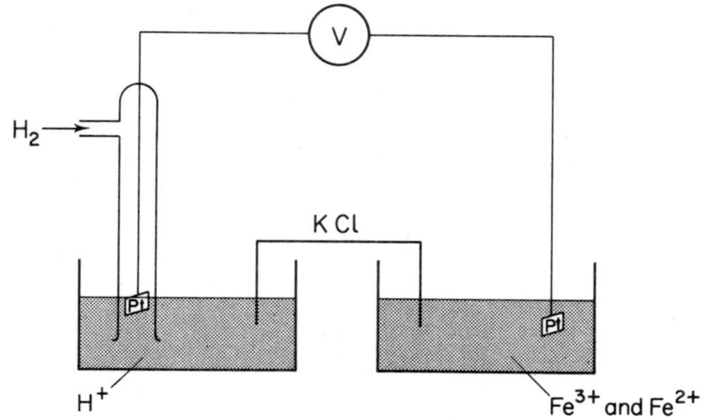

Fig. 1.2e. *Galvanic cell for the determination of sep of Fe^{3+}, Fe^{2+}*

Note that all species in aqueous solution should be at an activity of 1 mol dm^{-3} and the H_2 gas at a pressure of 101.325 kPa (1 atm).

As all species are in their standard state:

$$E^{\ominus}(\text{cell}) = E^{\ominus}(\text{Fe}^{3+}/\text{Fe}^{2+}) - E^{\ominus}(\text{she})$$

$$E^{\ominus}(\text{cell}) = +0.771 \text{ V}$$

As $E^{\ominus}$(cell) is positive, the cell reaction will proceed spontaneously in the direction written. By convention reduction occurs at the right hand electrode (a useful memory aid is that reduction and right keep the letter 'r' together!) and we write:

$$\text{Fe}^{3+} + \text{e} \rightarrow \text{Fe}^{2+}$$

Oxidation then occurs at the left hand electrode:

$$\text{H}_2 \rightarrow 2\text{H}^+ + 2\text{e}$$

Adding the two equations gives

$$H_2 + 2\,Fe^{3+} \rightarrow 2\,H^+ + 2\,Fe^{2+}$$

Note, that we multiply the iron(III)/iron(II) by two in order that the number of electrons is the same in both oxidation and reduction equations.

****

SAQ 1.2d

Write down the overall chemical equation for the cell represented by:

$$\text{Pt}\left|\begin{matrix}Fe^{3+} & Fe^{2+}\\ (a = 1\ \text{mol dm}^{-3}); & (a = 1\ \text{mol dm}^{-3})\end{matrix}\right\|\begin{matrix}Zn^{2+}\\ (a = 1\ \text{mol dm}^{-3})\end{matrix}\left|\text{Zn}\right.$$

What is the standard emf of this cell and in which direction will the reaction proceed?

Response

Oxidation occurs at left hand electrode, ie:

$$Fe^{2+} \rightarrow Fe^{3+} + e$$

and reduction at the right hand electrode, ie:

$$Zn^{2+} + 2e \rightarrow Zn$$

Thus the cell reaction is:

$$2\,Fe^{2+} + Zn^{2+} \rightarrow 2\,Fe^{3+} + Zn$$

The standard emf is:

$$E^{\ominus}(\text{cell}) = E^{\ominus}(\text{Zn}^{2+}/\text{Zn}) - E^{\ominus}(\text{Fe}^{3+}, \text{Fe}^{2+})$$

inserting the values from Fig. 1.2c give

$$E^{\ominus}(\text{cell}) = -0.7628 - 0.771$$

$$= -1.534 \text{ V}$$

Thus, the cell as written has a negative emf and the cell reaction as written is not spontaneous. The spontaneous reaction is:

$$\text{Zn} + 2\,\text{Fe}^{3+} \rightarrow \text{Zn}^{2+} + 2\,\text{Fe}^{2+}$$

ie zinc metal will reduce iron(III). The cell should be written in the reverse direction.

****

SAQ 1.2e

A galvanic cell consisting of a she (as the left hand electrode) and a rod of zinc dipping into a solution of Zn^{2+} ions at 298 K gave a measured emf of −0.789 V. What is the activity of the zinc ions?

Response

Using the Nernst equation:

$$E = \text{E}^{\ominus} + \frac{RT}{nF} \ln a(\text{Zn}^{2+})$$

From Fig. 1.2c

$$E^{\ominus}(\mathrm{Zn^{2+}/Zn}) = -0.7628\ \mathrm{V}$$

Then

$$-0.789 = -0.7628 + \frac{8.314 \times 298}{2 \times 96487} \ln a(\mathrm{Zn^{2+}})$$

$$\ln a(\mathrm{Zn^{2+}}) = (-0.026) \times \frac{2 \times 96487}{8.314 \times 298}$$

$$= -2.0407$$

$$a(\mathrm{Zn^{2+}}) = 0.130\ \mathrm{mol\ dm^{-3}}$$

****

SAQ 1.3a

Why is a liquid junction potential created between two solutions of hydrochloric acid of different concentrations?

Response

The H^+ ions migrate across the boundary, from the more concentrated to the more dilute solution at a faster rate than do the Cl^- ions. This results in a build up of more positive ions than negative ions on the dilute side of the boundary. This imbalance of charge results in the liquid junction potential.

****

SAQ 1.3b

Why are potassium chloride bridges used to minimise liquid junction potentials?

Response

The mobility of the K^+ and Cl^- ions are about the same and they diffuse across the boundary at virtually equal rates. Thus, there is little charge imbalance and minimal liquid junction potential.

****

SAQ 2.1a

Which of the following describes the Nernstian response of a chloride ion selective electrode at 298 K?

(*i*) $E(\text{cell}) = E' + 0.0257 \ln a(\text{Cl}^-)$

(*ii*) $E(\text{cell}) = E' - 0.0257 \ln a(\text{Cl}^-)$

(*iii*) $E(\text{cell}) = E' + 0.0128 \ln a(\text{Cl}^-)$

(*iv*) $E(\text{cell}) = E' - 0.0128 \ln a(\text{Cl}^-)$

(*v*) $E(\text{cell}) = E' \pm 0.0257 \ln a(\text{Cl}^-)$

Response

(*i*) Bad luck – you have calculated the value of the constant correctly, but the charge on the chloride ion is negative. The sign before the ln term should, therefore, be negative.

(*ii*) Good – correct answer. If you didn't get this answer I hope you now realise how it was obtained. The negative sign is chosen because the charge on the ion is negative. The constant is calculated as follows:

$$\frac{RT}{nF} = \frac{8.314 \times 298}{1 \times 96487} = 0.0257$$

(*iii*) You have made two common errors. Firstly, the charge on the ion is negative and the sign in the equation should also be negative. Secondly, the chloride ion has only one charge (Cl^- not Cl^{2-}) and n is, therefore, 1.

(*iv*) You have chosen the correct sign for the equation but the chloride ion has only one charge (Cl^- not Cl^{2-}) and n is, therefore, 1.

(*v*) You have calculated the value of the constant correctly, but you need to select the negative value from the ± sign. The ± sign refers to a general case and if the ion has a negative charge the sign must match it. Similarly, if the ion was positively charged, you would need to select the positive sign.

****

SAQ 2.1b

Draw a cell for the estimation of copper(II) ions by potentiometry.

Response

Your diagram should resemble Fig. 2.1b.

One point to note is that the indicator electrode should be an ise which responds to copper(II) ions. This will be a specially constructed electrode and not a rod of pure copper, because metallic copper also responds to ions other than Cu^{2+}. Metallic copper is therefore, not a specific or even selective electrode.

****

SAQ 2.2a

List two reasons why concentration standards are used in preference to activity standards when calibrating ion-selective electrodes. What conditions must exist when using concentration standards?

Response

(*i*) it is easier to prepare standard solutions of known concentration;

(*ii*) it facilitates comparison with other techniques.

Conditions

An isa must be used or the calibration line will be curved. The use of an isa generally poses few problems, though if we are dealing with samples covering a large concentration range, the approximation in the derivation of the relationship:

$$I(\text{mixture}) = I(\text{isa})$$

does not hold. This is because that at higher sample concentrations I(isa) is not very much greater than I(sample).

****

SAQ 2.2b

Select the appropriate phrase which best completes the sentence 'The purpose of adding an ionic strength adjustor to each sample and standard before each measurement is taken is to ...'

(*i*) ensure that all solutions have the same total ionic strength;

(*ii*) adjust the ionic strength to 1 mol dm^{-3};

(*iii*) allow emf measurements to be related directly to the activity of the ion under examination;

(*iv*) increase the measured emf giving more accurate readings.

Response

(*i*) Correct answer.

(*ii*) There is no requirement to know the actual ionic strength. The only requirement is that it is identical for all standards and sample.

(*iii*) The emf will always be related to the activity, whether an isa is present or not. The isa allows a linear relationship between log (concentration) and emf.

(*iv*) Whilst the emf will change on addition of an isa, all standards and samples will increase by the same amount. Consequently, there is no increase in accuracy as such.

****

SAQ 2.3a

A student was calibrating a calcium ise and obtained the following results. What is the concentration of a sample (marked S) which was measured at the same time?

Concentration of Ca^{2+} /mol dm^{-3}	emf /mV
1.00×10^{-4}	-2
5.00×10^{-4}	$+16$
1.00×10^{-3}	$+25$
5.00×10^{-3}	$+43$
1.00×10^{-2}	$+51$
S	$+33$

Response

I hope you found the concentration of the sample to be 2.1×10^{-3} mol dm^{-3}. The calibration graph is shown in Fig. 2.3c.

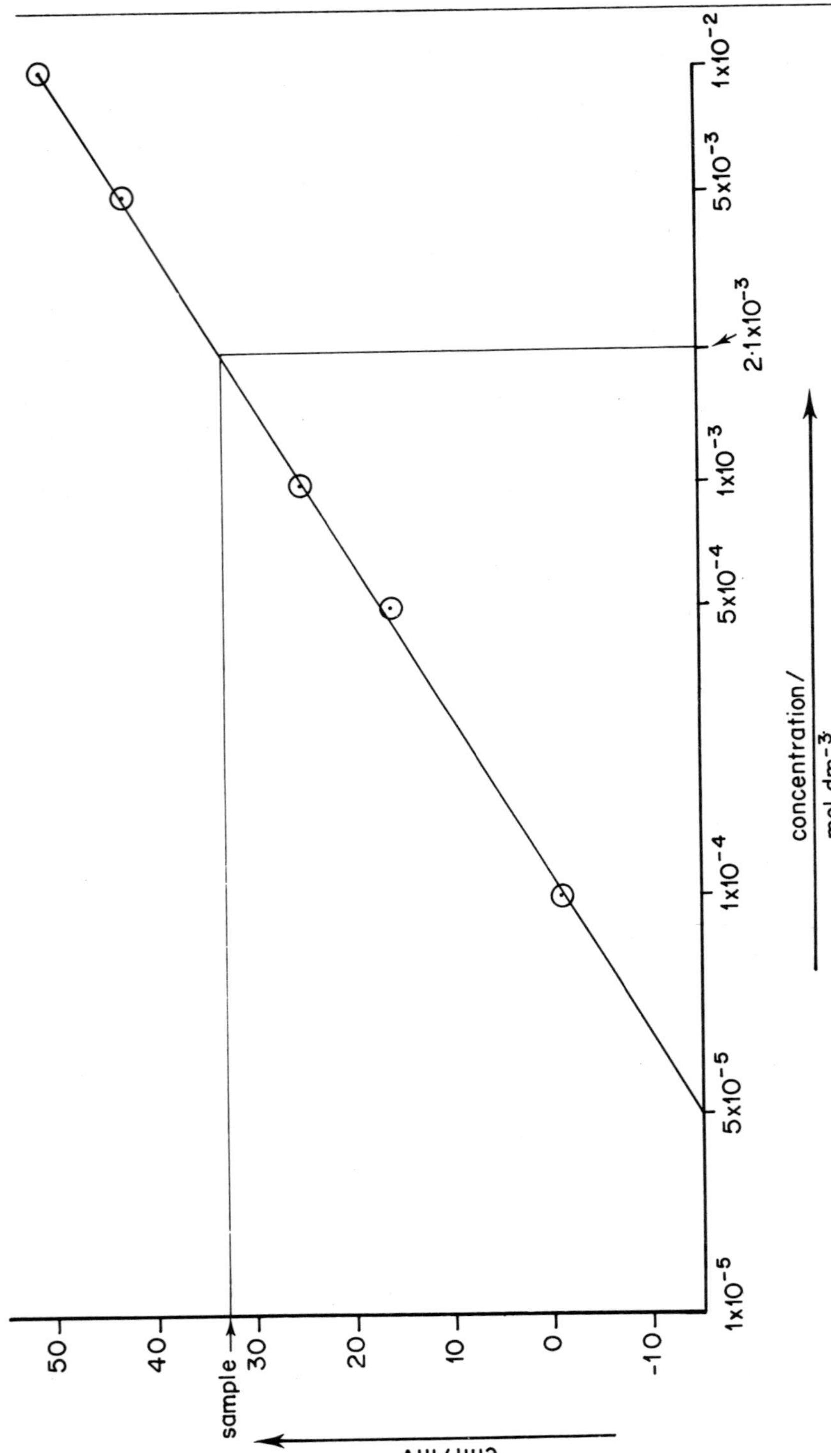

Fig. 2.3c. *Calibration curve for calcium ise*

SAQ 2.3b

Which of the following is *not* a reason why concentration standards are generally used in potentiometry?

(*i*) Concentration standards may be prepared from a primary standard.

(*ii*) The calibration graph of log (concentration) against emf is always a straight line.

(*iii*) Concentration values facilitate comparison with other techniques.

(*iv*) Accurate activity standards are difficult to prepare as values of the activity coefficient are the result of assumptions.

Response

The correct answer to this question is (*ii*).

The calibration graph of log (concentration) against emf is only linear at ionic strengths less than 10^{-4} mol dm^{-3} or when an isa is present. In other cases it is a curve.

If you chose one of the other responses, you selected one of the reasons why concentration standards are chosen, whilst the question asks why they are not chosen.

****

SAQ 2.3c

> Use Fig. 2.3a to estimate the activity of the potassium ion in a 0.200 mol dm^{-3} solution of potassium chloride.

Response

The correct answer is 0.145 mol dm^{-3}.

From the table we have:

$$\gamma(K^+) = 0.727$$

The activity is then calculated using:

$$a(K^+) = \gamma(K^+)\ c(K^+)$$

$$= 0.727 \times 0.200$$

$$= 0.145 \text{ mol dm}^{-3}$$

****

SAQ 2.4a

> Calculate the maximum tolerated activity of silver ions for a 10% error when using a potassium ise to measure K^+ ions in the range 10^{-3} to 10^{-4} mol dm^{-3}. $k_{K,Ag} = 1.00 \times 10^{-4}$.

Response

The maximum interference will occur with the most dilute solution, ie 10^{-4} mol dm^{-3}. Using Eq. 2.4c we have:

$$\frac{a(K^+)}{a(Ag^+)} = \frac{k}{\text{error}} \times 100$$

$$a(Ag^+) = \frac{\text{error} \times a(K^+)}{k \times 100} = \frac{10 \times 10^{-4}}{10^{-4} \times 100}$$

$$= 0.100 \text{ mol dm}^{-3}$$

To keep the error in measurement below 10%, we must keep the activity of the silver ion below 0.100 mol dm^{-3}.

****

SAQ 2.4b

Examine each of the following statements and decide whether it is true or false.

(*i*) Ion selective electrodes respond to ions other than the ion under investigation.

(*ii*) A large value for the selectivity coefficient indicates a highly interfering ion.

(*iii*) Selectivity coefficients are constant across the entire concentration range.

(*iv*) Selectivity coefficient is the ratio of an electrode's response to an interfering ion relative to the ion under investigation.

Response

(*i*) This statement is certainly true. It is this response to other ions leads to interference when using ises.

(*ii*) This statement is also true. The selectivity coefficient is the ratio of how the electrode responds to the interfering ion, compared to how it responds to the ion under investigation.

(*iii*) This statement is false. A lot of users forget the selectivity coefficients vary with the total ionic strength and these values should only be used for guidance.

(*iv*) This statement is also true. In fact it is the definition of selectivity coefficient.

****

SAQ 2.5a

Fig. 2.5c is a calibration graph for an Orion nitrate (NO_3^-) ise.

Indicate on the graph:

(*i*) the limit of Nernstian response;

(*ii*) the lower limit of measurement as defined by IUPAC. ⟶

SAQ 2.5a (cont.)

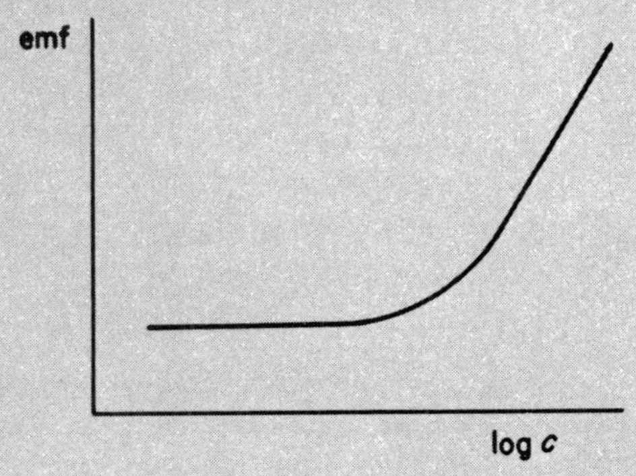

Fig. 2.5c. *Calibration graph for nitrate ise*

Response

(*i*) The lower limit of Nernstian response is the point of deviation from linearity and is denoted by A in the diagram.

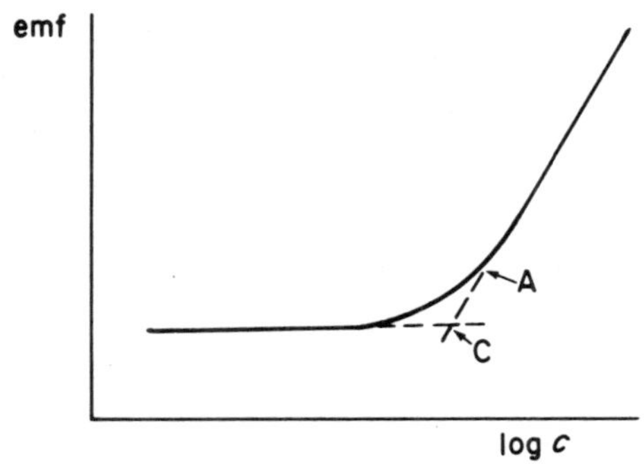

Fig. 2.5c - extended

(*ii*) The IUPAC limit is obtained by extrapolation as shown in the diagram. The lower limit of response is indicated by point C.

****

SAQ 2.5b

Which of the following factors influence the response time of an ise?

(*i*) temperature;
(*ii*) interfering ions;
(*iii*) the voltmeter;
(*iv*) type of electrode;
(*v*) stirring rate.

Response

The simple answer to this one is that all of these factors influence the response time.

(*i*) An increase in temperature reduces the response time.

(*ii*) Interfering ions increase the response time.

(*iii*) The damping of the voltmeter display increases response time.

(*iv*) Solid state and glass electrodes respond more quickly than do liquid ion exchange and gas sensing probes.

(*v*) The faster the stirring rate, the shorter the response time.

****

SAQ 2.6a

Which of the following statements are true?

(*i*) The direct reading method is quick once the electrode is calibrated.

(*ii*) The accuracy of analysis using ion-selective electrodes is the same for singly and doubly charged ions.

(*iii*) Calibration of ion selective electrodes requires a minimum of two standards.

(*iv*) A pH meter is in effect a voltmeter calibrated in pH units. A decrease in measured emf of 59.1 mV always corresponds to an increase in pH of one unit.

(*v*) pH meters should always be calibrated in the same pH range as that of the sample(s).

Response

(*i*) True – this is the major advantage of the method and sample throughputs up to 120 per hour are possible.

(*ii*) False – the slope of the calibration plot for singly charged ions is 59.1 mV per decade at 298 K, whilst that for doubly charged ions is 29.6 mV per decade at the same temperature. The former calibration graph is, therefore steeper than the latter and this leads to greater accuracy of analysis. In practice a good analyst should obtain accuracies in the region of $\pm 2\%$ for singly charged and $\pm 4\%$ for doubly charged ions.

(*iii*) False – if we assume that the electrode has a Nernstian response it is possible to calibrate the electrode using a single point. This is often done with pH electrodes where a buffer solution of known pH is used. However, it should be remembered that the more points that are used to calibrate the electrode, the more accurate the analysis.

(*iv*) False – pH meters are certainly voltmeters calibrated in pH units. However the slope of 59.1 mV per decade only applies to electrodes having Nernstian response at 298 K. At temperatures other than 298 K, the slope will be different and most meters have either a temperature control or automatic compensation to alter the slope to match the temperature.

(*v*) True – very few electrode systems exhibit linear calibration over very wide concentration ranges. It is, therefore, advisable to calibrate in the range in which you are working.

****

SAQ 2.6b

A fluoride ion selective electrode was used to measure the concentration of F^- in a cup of tea. When immersed in a mixture of 25 cm^3 of tea and 25 cm^3 of ionic strength adjustor, the electrode gave a reading of 98 mV. When 2.0 cm^3 of a 100 ppm solution of F^- was added to this mixture, the measured emf dropped to 73 mV. Calculate the concentration of fluoride ions in the tea.

Response

Substituting in Eq. 2.6e gives:

$$c = \frac{100 \times 2}{(50 + 2)\,\text{antilog}\,[(73 - 98/-59.1)] - 50}$$

$$c = 2.27$$

But the original solution of tea was diluted by a factor of two when the isa was added. Therefore the concentration of fluoride ions in tea = 4.5 ppm

****

SAQ 2.6c

50 cm^3 of a solution of Cu^{2+} was analysed using a multiple standard addition method. When incremental amounts of a standard 0.100 mol dm^{-3} solution of Cu^{2+} were added to the sample, the following emf readings were obtained.

Volume of addition /cm^3	emf /mV
0	99.8
1.00	102.5
2.00	104.6
3.00	106.3
4.00	107.9

→

SAQ 2.6c (cont.)

> A blank solution gave an emf reading of 70.0 mV.
>
> Estimate the concentration of the copper(II) ions in the original solution. (If Gran's plot paper is available use this.)

Response

When using ordinary graph paper, we must calculate antilog $[E(\text{cell})/S]$ for the above data. As copper(II) is a doubly charged, positive ion, the slope will be 29.1 mV.

Volume of addition /cm^3	emf /mV	antilog E/S
0	99.8	2688
1.00	102.5	3329
2.00	104.6	3931
3.00	106.3	4497
4.00	107.9	5104

The graph of antilog (E/S) against volume is illustrated below and extrapolation to the abcissa gives an intercept at -4.6 cm^3.

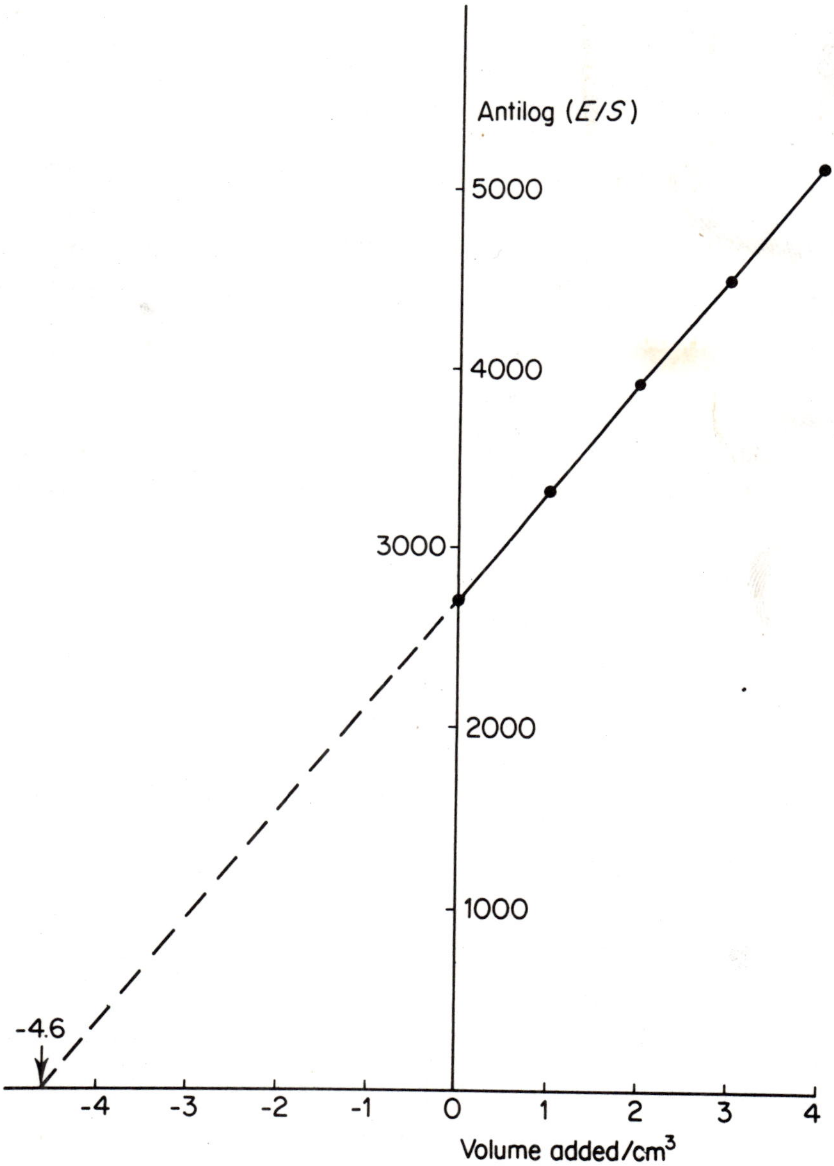

Fig. 2.6j. *Plot of antilog (E/S) against volume*

Substituting in Eq. 2.6f gives:

$$c = 4.6 \times 0.100/50$$

$$= 9.2 \times 10^{-3} \text{ cm}^3$$

If we use non-volume corrected Gran's plot paper, we can plot the emf values directly onto the ordinate. For a doubly charged ion, each major division on the paper corresponds to 2.9 mV. We also plot the volume axis at the emf of the blank, ie 70.0 mV.

The Gran's plot is shown in Fig. 2.6k and by extrapolation we obtain an intercept at -4.6 cm^3. The remainder of the calculation is identical to the method using conventional graph paper.

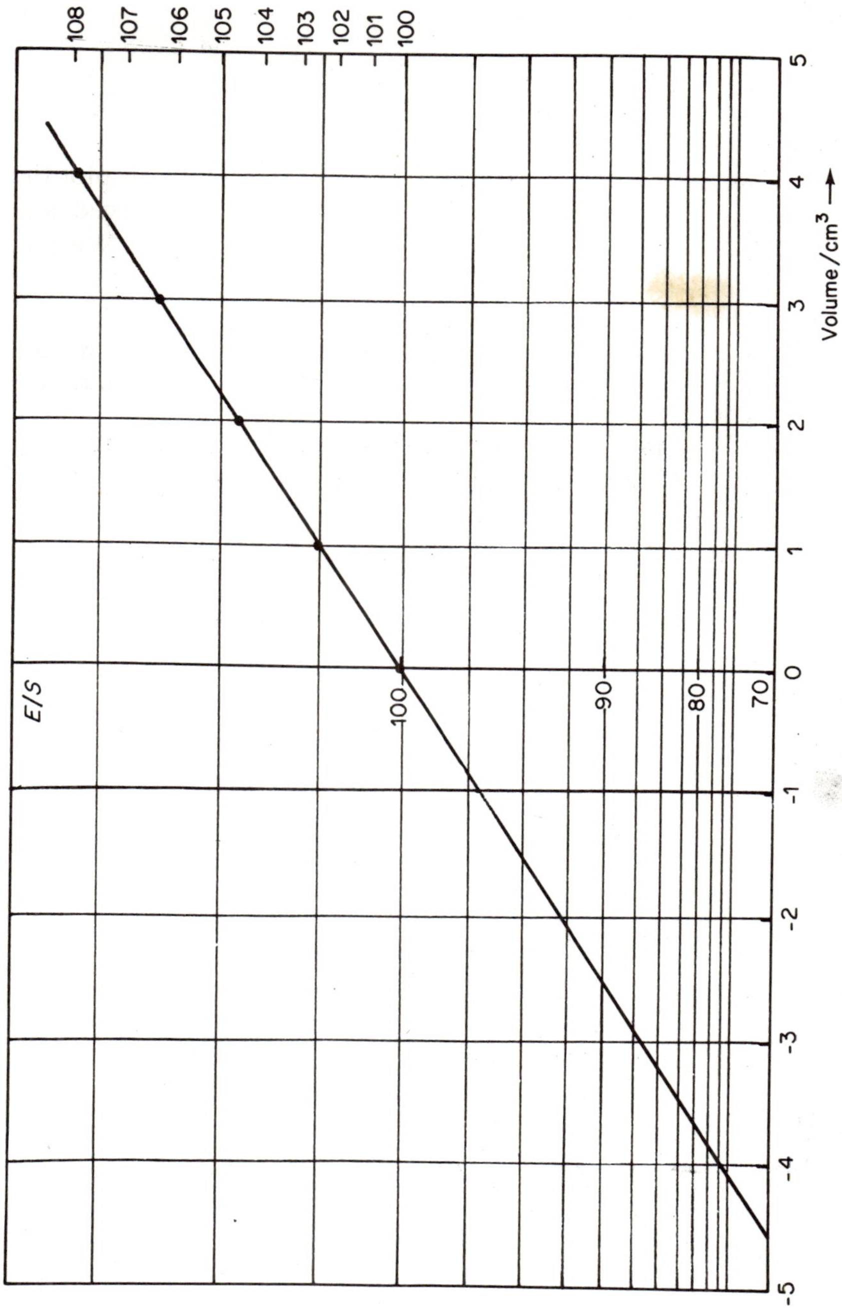

Fig. 2.6k. *Gran's plot of (E/S) against volume*

SAQ 2.6d

A student calibrated a Ca^{2+} ion selective electrode using two standard solutions at constant ionic strength and 298 K. He obtained the following results

Ca^{2+} concentration /mol dm^{-3}	emf /mV
1.00×10^{-3}	142
1.00×10^{-4}	113

What is the concentration of a test solution which gave an emf reading of 125 mV under the same conditions?

Response

We construct a calibration graph using semi-log graph paper as illustrated below:

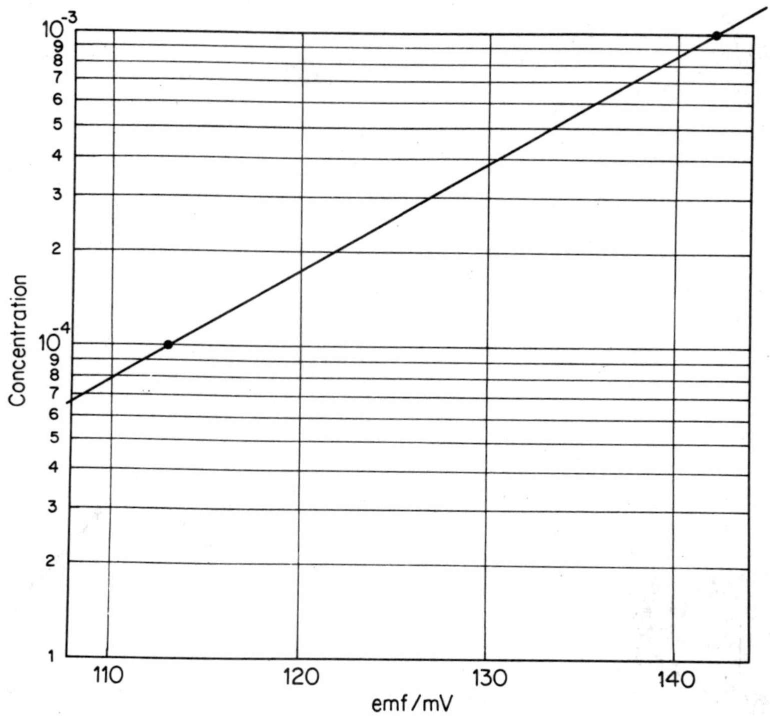

Fig. 2.61. *Calibration graph on semi-log paper*

The concentration can be read off directly from the axis.

Concentration of calcium(II) is 2.6×10^{-4} mol dm^{-3}.

****

SAQ 2.6e

A 50.0 cm^3 sample of a lead(II) solution was analysed using a standard addition method with a lead(II) ion selective electrode. The initial potential was 302 mV, which rose to 350 mV on addition of 1.00 cm^3 of a 1.00 mol dm^{-3} solution of lead(II). Assuming that all measurements were made at constant ionic strength and 298 K, calculate the concentration of the original solution.

Response

We do not need to worry about dilution effects on the sample as it is less than 2%. The standard, however is diluted by a factor of 50 during the addition. If we now use Eq. 2.6d

$$c = x/\{\text{antilog}[(E_2 - E_1)/\pm S] - 1\}$$

Substituting gives:

$$c = (1.00/50)/\{\text{antilog}[(350 - 302)/29.6] - 1\}$$

$$= 4.90 \times 10^{-4} \text{ mol dm}^{-3}$$

Remember that for the Cd^{2+} the electrode slope is +29.6 mV

****

SAQ 2.6f

In the potentiometric titration of 100 cm^3 of a chloride solution with a 2.00×10^{-3} mol dm^{-3} solution of silver nitrate, the emf was measured using an Ag^+ ise-reference assembly. The following results were obtained.

Volume $AgNO_3$ /cm^3	emf/mV blank	titration
1.00	278	
1.50	287	
2.00	294	
2.50	299	262
3.00	304	273
3.50		282
4.00		288
4.50		294
5.00		297
5.50		300
6.00		304

Using the Gran's plot method determine the equivalence point of the titration and the concentration of the chloride solution. If you have access to 10 % volume corrected Gran's plot paper use this and take the blank into consideration. If not plot antilog(E/S) against V (and ignore the blank readings).

Response

(*i*) Plot of antilog (E/S) against V (Fig. 2.6m). The following table is first constructed. $S = 59.1$ mV

Volume	E/mV	E/S	antilog(E/S)
2.5	362	4.42	27112
3.0	273	4.60	41618
3.5	282	4.77	59098
4.0	288	4.87	74661
4.5	294	4.97	94323
5.0	297	5.02	106018
5.5	300	5.076	119163
6.0	304	5.14	139259

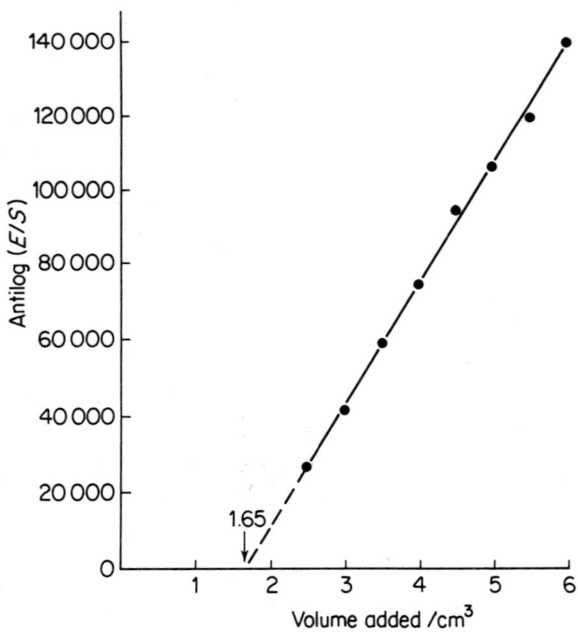

Fig. 2.6m. *Graph of antilog (E/S) against volume*

The intercept is 1.65 cm^3, therefore the concentration of chloride was as given by:

$$2.00 \times 10^{-3} \times 1.65/100$$

$$= 3.3 \times 10^{-5} \text{ mol dm}^{-3}$$

(*ii*) Using 10 % volume corrected Gran's plot paper. We firstly draw the ordinate which represents the emf values.

We then plot the points for the blank titration and extrapolate the line through these points back to the ordinate. We now draw the abcissa at the intercept of this line with the ordinate.

Having done this, we plot the data for the titration. Note, that as we are monitoring Ag^+ ions, we are interested in the increase in emf after equivalence. We now extrapolate the titration values back to intercept with the abcissa.

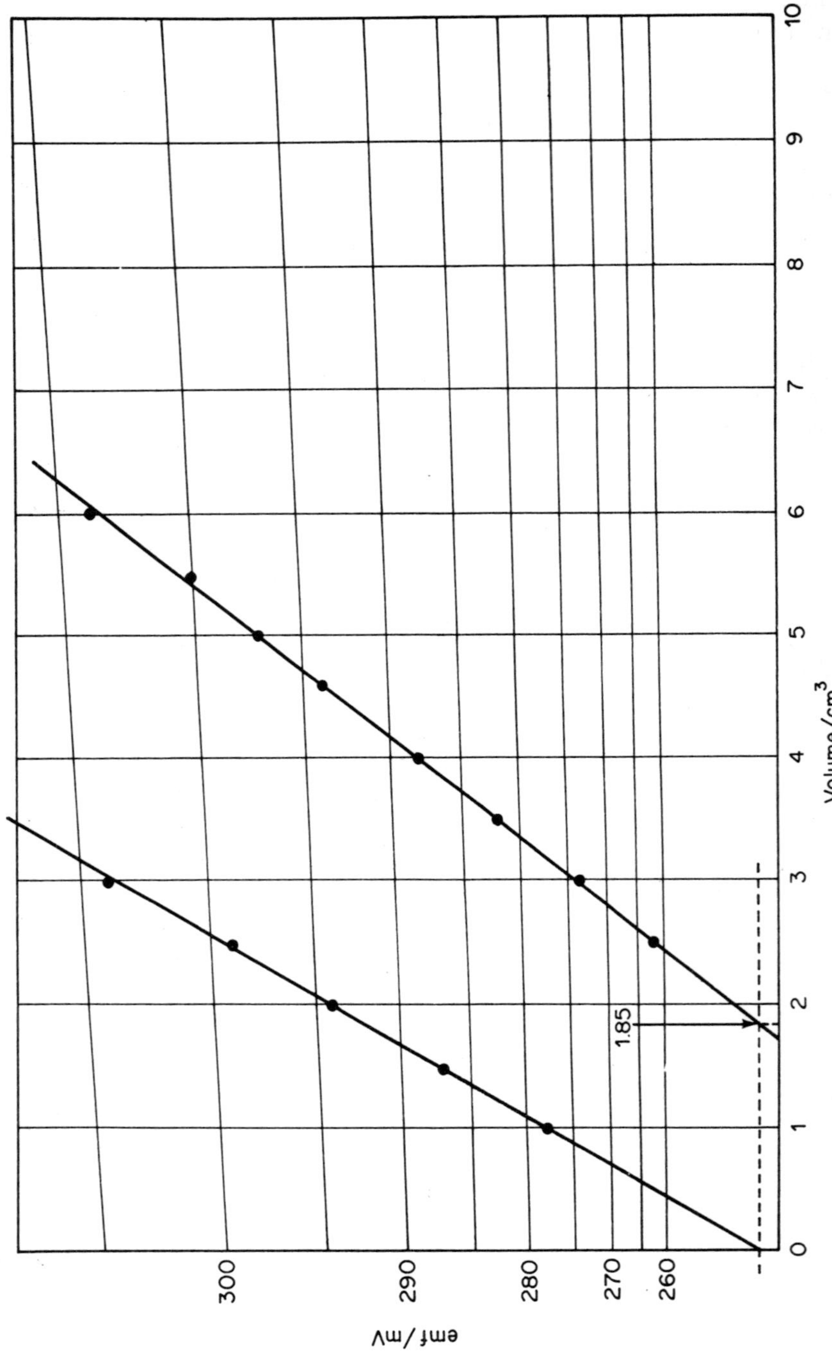

Fig. 2.6n. *Gran's plot*

From the graph, the intercept is 1.85 cm^3. Therefore, at equivalence 1.85 cm^3 of a 2.00×10^{-3} mol dm^{-3} solution of silver nitrate.

As Ag^+ and Cl^- react with 1 : 1 stoichiometry

$$\therefore \quad \text{concentration of } Cl^- \text{ ions} = \frac{2.00 \times 10^{-3} \times 1.85}{100}$$

$$= 3.70 \times 10^{-5} \text{ mol dm}^{-3}$$

The slight difference in the two methods is due to the drawing of the best line and the use of volume correction when using the Gran's plot paper.

****

SAQ 2.7a

Select reference electrodes for the following applications, giving reasons for your choice.

(*i*) For a combined glass electrode to measure pH.

(*ii*) For use with a Cl^- ise at 298 K.

(*iii*) For use with a potassium ise to analyse sea water.

(*iv*) For use with an F^- ise to examine drinking water.

Response

(*i*) This is a common application of potentiometry and we will need a small, cheap electrode. The silver,silver chloride electrode will fit the bill ideally. Fig. 2.7b illustrates such a combined electrode.

(*ii*) I hope you remembered that both the sce and Ag,AgCl electrodes contain KCl in the bridge solution. This would leak out and contaminate our test solution and standards. To avoid the complication of a double junction, the best reference electrode is the mercury,mercury(I) sulphate(VI).

(*iii*) As in the previous example, contamination of the test solution would occur because the K^+ ions in the bridge of each of the electrodes would diffuse into the sample. We could either use one of the electrodes described above, but change the KCl solution for an NaCl solution of similar concentration. Alternatively, we could use an Ag,AgCl double junction electrode. The latter is best as it is easier to refill the outer solution of the double junction with NaCl, than completely strip down and refill a single junction electrode.

(*iv*) As the chloride ion is similar to the fluoride ion, it is advisable to check the fluoride electrode's selectivity coefficient. Fortunately, Cl^- does not interfere with the electrode's operation to a significant extent. We can, therefore, use the Ag,AgCl reference.

****

SAQ 2.8a

Identify the factors which you think are important in selecting:

(*i*) an indicator electrode;
(*ii*) a reference electrode and
(*iii*) a method,

for the routine potentiometric analysis of drinking water for calcium and magnesium ions (ie water hardness).

Response

(*i*) choice of electrode

In this analysis, we need to measure two divalent cations simultaneously. Fortunately, it is possible to obtain an electrode which responds equally to both calcium and magnesium cations. You could describe the electrode as a calcium ise which has a selectivity coefficient of unity for the magnesium ion. In drinking water, the concentration of other divalent cations is likely to be small and there is little likelihood of interference.

(*ii*) choice of reference electrode

None of our common reference electrodes contain divalent cations and there is little chance of contamination. All samples and standards are likely to be at room temperature and there is no need to examine temperature coefficients. The only possible problem could be if the mercury(I)sulphate(VI) reference was used. Calcium sulphate(VI) is limitedly soluble and a precipitate of $CaSO_4$ could form at the interface of the reference electrode and the solution. It is, therefore, sensible to use an sce or an Ag,AgCl electrode.

(*iii*) choice of method

For routine analysis where there are likely to be a number of samples, the most satisfactory procedure is the direct reading method. There are not likely to be any major interference problems and there is no requirement to use the more time consuming methods such as multiple standard addition.

****

SAQ 3.1a

Sketch and label a conventional glass electrode.

Response

Your sketch should resemble that shown in Fig. 3.1a.

****

SAQ 3.1b

Which of the following descriptions best describes the term 'boundary potential'?

(*i*) The effect of the hydration of the surface layers of the glass.

(*ii*) The effect of the difference in activity between the H^+ ions in the test solution and the hydrated layer.

(*iii*) The effect of the difference in the structure of the inner and outer glass surfaces.

(*iv*) The effect of the difference between the activity of the H^+ ions in the test and reference solutions.

Response

(*i*) This is not the answer to the question – the hydration of the surface layer is only the first step in establishing the boundary potential. Once established, it is the difference in activity between the H^+ in the solution and the hydrated layer that results in the boundary potential.

(*ii*) Correct answer, if the activity of the H^+ ions in the solution is different to that in the hydrated layer, a potential difference is set up. This potential difference is the boundary potential.

(*iii*) Incorrect – if the inner and outer glass surfaces are different, a potential will exist across the membrane; this is called the assymetry potential. The correct answer to this question is (*ii*). If the activity of the H^+ ions in the solution is different to that in the hydrated layer, a potential difference is set up. This potential difference is the boundary potential.

(*iv*) This is not the answer to this particular question. The difference between the activities of the H^+ in the test and reference solutions certainly affects the overall measured potential but not the individual boundary potentials. The correct answer to this question is (*ii*). If the activity of the H^+ ions in the solution is different to that in the hydrated layer, a potential difference is set up. This potential difference is the boundary potential.

****

SAQ 3.1c

Which of the following circumstances would result in a change in the assymetry potential (more than one of the list may contribute)?

(*i*) The electrode has been used to measure the pH of solutions in which the activity of the H^+ ions in the test solution and the inner reference solution are different.

(*ii*) The electrode has not been buffered for over a day.

(*iii*) The electrode has been stored in a cardboard box in a cupboard.

(*iv*) The electrode has been used to measure the pH of solutions containing sodium hydroxide.

Response

The two factors that affect the assymetry potential are (*iii*) and (*iv*). The assymetry potential is that which exists across the membrane when the activity of H^+ in the test and reference solutions are the same. There are a number of reasons for it and if you chose (*i*) or (*ii*). You check in the text for them.

(*i*) This does not affect the assymetry potential, though it does affect the overall cell potential.

(*ii*) Buffering the electrode compensates for the assymetry potential, rather than causing it. It is advisable that glass electrodes are buffered (calibrated) at least once a day when in regular use.

(*iii*) This is one of the commonest reasons for large assymetry potentials. When an electrode is stored dry, the outer hydrated layer 'dries out' and the activity of the H^+ ions decreases. When the electrode is next placed in a solution and the potential measured, considerable drift will occur as the outer surface becomes hydrated again. It is, therefore, advisable that glass electrodes be stored wet.

(*iv*) Sodium hydroxide has the property of etching glass and it will affect the composition of the outer surface of the membrane. This, in turn, will affect the degree of hydration and the activity of the H^+ ions in the hydrated layer. The assymetry potential will, therefore, change relatively slowly with repeated measurement.

****

SAQ 3.1d

Which of the following would cause an alkaline error?

(*i*) High activities of Na^+.

(*ii*) Low activities of H^+.

(*iii*) The presence of a non-aqueous solvent.

(*iv*) The glass membrane contains Li_2O.

Response

(*i*) Correct answer. The Nernst equation for the glass electrode in the presence of interfering ions can be written (Eq. 3.1c)

$$E(\text{cell}) = E^* + (RT/F) \ln [a(\text{H}^+) + k_{\text{H,Na}} a(\text{Na}^+)]$$

where $k_{H,Na}$ is the the selectivity coefficient of the electrode for H^+ over Na^+. If the activity of Na^+ is high, then it affects E(cell) in a similar way to H^+. This gives a lower or more acid pH. The term alkaline error is used because most strongly alkaline solutions contain Na^+.

(*ii*) Incorrect. A low H^+ activity in itself will not cause alkaline error, although the pH will be alkaline.

(*iii*) This will cause acid error as it affects the activity of the water.

(*iv*) Incorrect. This will minimise alkaline error. For glass electrodes containing Li_2O, the value of the selectivity coefficient ($k_{H,Na}$) will be very much smaller.

****

SAQ 3.1e

> List some factors which influence the accuracy of pH measurement.

Response

There are a number of factors you need to consider.

(*a*) Errors due to calibration.

Buffer solutions cannot be prepared more accurately than ±0.01 pH units. We cannot calibrate the electrode better than this. Further, errors will occur if the electrode is calibrated at a value outside the range over which it will operate and at a temperature very different from that of the test solution.

(*b*) Errors due to test solution.

The common errors to be considered here are acid and alkaline error which occur at the extremes of pH measurement. The former is due to the effect on the water of a high concentration of any ion or a non-aqueous solvent. The latter is a general term for the effect of an interfering ion on the electrode.

Other errors due to differences in the composition of test solutions are due to liquid junction potentials between the solution and the reference electrode.

(*c*) Errors due to equipment.

As with all analytical measurements, random errors will be present due to fluctuations in the power supply to the meter, parallax errors in reading analogue scales etc.

****

SAQ 3.1f

What effect would changing the composition of the glass from Na_2O/SiO_2 to $Li_2O/Al_2O_3/SiO_2$ have on a conventional glass electrode?

(*i*) Make the membrane more robust.

(*ii*) Make the electrode more selective towards H^+.

(*iii*) Make the electrode more selective towards Na^+.

(*iv*) Convert the electrode into an aluminium ise.

Response

(*i*) Unfortunately, this is the wrong answer. Glass electrodes are very fragile and anything that would make it more robust would be welcomed.

(*ii*) Wrong answer – the change in the composition of the glass makes it *less* selective towards H^+ and *more* selective towards Na^+.

(*iii*) Correct answer – the change in the composition of the glass makes it *less* selective towards H^+ and *more* selective towards Na^+

(*iv*) Incorrect – the change in the composition of the glass makes it *less* selective towards H^+ and *more* selective towards Na^+.

****

SAQ 3.1g

Which of the following ions would produce the greatest interference on a potassium glass electrode?

(*i*) Ag^+;

(*ii*) NH_4^+;

(*iii*) Na^+;

(*iv*) Ca^{2+}.

Response

The correct answer is (*i*).

The order of response for a potassium glass electrode is:

$$H^+ > Ag^+ > K^+ = NH_4^+ > Na^+ > Li^+ \ldots \gg Ca^{2+}.$$

Thus, the silver ion has by far the greatest level of interference when measuring K^+. If Ag^+ ions were present, they would have to be removed by precipitation or complexation.

****

SAQ 3.2a

> **Sketch an all solid state ise using a silver sulphide membrane.**

Response

Your electrode should resemble Fig. 3.2f.

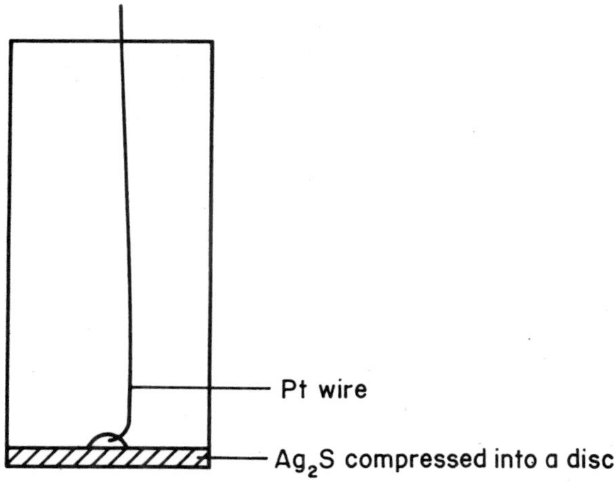

Fig. 3.2f. *Solid state ise with silver sulphide membrane*

Note that the membrane is a compressed disc of Ag_2S, rather than a single crystal, which is difficult to grow. Some inert binder, such as PVC or silicone rubber, can also be added to the disc to improve its strength.

****

SAQ 3.2b List three factors which contribute towards the selectivity of solid state membranes.

Response

The selectivity of the solid-state membranes relies on an ionic equilibrium at the membrane surface. Only ions of

– the same size,

– the same shape, and

– the same charge,

as the ions in the membrane can occupy the lattice sites and take part in the equilibrium.

Only rarely have two different ions the same combination of these three parameters and this makes the electrodes highly selective.

****

SAQ 3.2c

Which of the following could be used as a solid-state membrane for an I^- ise?

(*i*) sodium iodide;

(*ii*) silver iodate;

(*iii*) iodoform;

(*iv*) silver iodide;

(*v*) lead(II)iodide.

Response

In selecting the correct answer, we need to find an insoluble crystal which contains I^- ions. The only one of the list which satisfies these two factors is (*iv*) – silver iodide, though (*v*) – lead iodide goes a long way towards satisfying the criteria.

(*i*) Sodium iodide would make a poor membrane as it is very soluble and it would dissolve in the test solution.

(*ii*) Silver iodate has the formula $AgIO_3$ and contains Ag^+ and IO_3^- ions. Hence it could act as an ise for either of these two ions, but not for I^-.

(*iii*) Iodoform is certainly insoluble and contains iodine. However, the iodine is covalently bonded and not in the form of I^-. Hence it cannot be used as an ise membrane.

(*iv*) Silver iodide can certainly be used as a membrane because it satisfies the criteria of being insoluble and containing I^- ions.

(*v*) Lead iodide is certainly a possible membrane as it satisfies the criteria of being insoluble and containing iodide ions at room temperature. Unfortunately, its solubility increases substantially with temperature. Its effectiveness as a membrane, consequently, decreases.

****

SAQ 3.2d

Silver sulphide is preferred to silver chloride when constructing a membrane for an Ag^+ ise because:

(*i*) it contains more silver ions;

(*ii*) its solubility product is lower;

(*iii*) chloride ions interfere;

(*iv*) it forms a stronger disc.

Response

(*i*) There are certainly more silver ions in the Ag_2S crystal, and whilst this helps the ionic conduction in the membrane, it is not the main reasons why Ag_2S is used.

(*ii*) Correct answer, the lower the solubility product of the membrane material, proportionally less of it must dissolve to establish the equilibrium. The response time and lifetime of the Ag_2S membrane will be very much better than those of the AgCl.

(*iii*) Cl^- ions in the membrane will not affect the electrodes ability to monitor Ag^+ in solution. If, however, free Cl^- ions were present in the test solution, then they would precipitate out the Ag^+. The Ag^+ ise responds to free silver ions in solution and if they were precipitated out, they would not be measured.

(*iv*) This statement is true if we consider homogeneous membranes. However, the addition of binding materials can strengthen a weak disc and heterogeneous discs would be of comparable strength. The criteria for selecting a membrane material would firstly be selectivity, lower limit of detection and response time, rather than strength.

****

SAQ 3.2e

The solubilities (mol dm^{-3}) of some common sulphides are:

arsenic(III)	1×10^{-5}
cadmium(II)	2×10^{-6}
copper(II)	2×10^{-8}
germanium(II)	3×10^{-2}
iron(II)	1×10^{-4}
lead(II)	2×10^{-6}
manganese(II)	9×10^{-5}
tin(II)	2×10^{-7}

Predict the order which the following cations will interfere with the operation of a Pb^{2+} ise using a PbS/Ag_2S membrane.

As^{3+} Cd^{2+} Cu^{2+} Ge^{2+} Fe^{2+} Mn^{2+} Sn^{2+}

Response

The interference is likely to be in the order of solubility. A cation that forms a more insoluble sulphide than PbS, will interefere strongly, whilst one with a smaller solubility will interfere less. The likely order of response of the Pb^{2+} ise is, therefore:

$$Cu^{2+} > Sn^{2+} > Cd^{2+} = Pb^{2+} > As^{3+} > Mn^{2+} > Fe^{2+} \gg Ge^{2+}$$

Note

This order does not take into account the kinetics of the interfering surface reaction. It may be that Pb^{2+} will move up the order when this is also taken into account. If you want to consider the likely level of interference, it is better to consult the values of selectivity coefficients provided by the manufacturer.

****

SAQ 3.2f

Are the following statements true or false when applied to solid-state membrane electrodes?

(*i*) They have relatively fast response times.

(*ii*) They should be stored in concentrated solutions of the ion for which they are selective.

(*iii*) Their lower limit of measurement is 10^{-5} mol dm^{-3}.

(*iv*) The lower limit of measurement is independent of pH.

Response

(*i*) This statement is true. Solid state membrane electrodes have the fastest response times of all ion selective electrodes. This is due to the rapid establishment of the boundary potentials and the fast transport of ions, *via* the defects, across the membrane.

(*ii*) This depends on the membrane, LaF_3 and Ag_2S membranes can be 're-charged' by immersion in concentrated solutions of the analyte ion. For example, a concentrated solution of F^- will help reverse the equilibrium:

$$LaF_3(s) + 3\,OH^-(aq) = La(OH)_3(s) + 3\,F^-(aq)$$

and remove the surface coating of $La(OH)_3$. On the other hand, silver halide electrodes react with excess of halide ion to form haloargentate ions:

$$AgCl(s) + Cl^-(aq) = AgCl_2^-(ag)$$

and the membrane would dissolve.

(*iii*) This is only true for the Cl^- ise using an AgCl membrane. Other electrodes will have different lower limits, depending on the solubility of the membrane.

(*iv*) This statement is false. High (alkaline) pH values may well result in the precipitation of metal cations such as Cu^{2+}. Alternatively, high pH values can result in contamination of the surface of the membrane, as with the LaF_3 membrane:

$$LaF_3(s) + 3\,OH^-(aq) = La(OH)_3(s) + 3\,F^-(aq)$$

Low (acid) pH values can also cause a reaction with the membrane. For example, Ag_2S reacts in acid solution to give off H_2S gas:

$$Ag_2S(s) + 2\,H^+(aq) = H_2S(g) + 2\,Ag^+(aq)$$

This effectively increases the solubility of the membrane, reducing the lower limit of measurement.

Solid state membrane electrodes, therefore, have an optimum working range and are generally used with buffer solutions. For concentration measurements, the buffer is incorporated into the ionic strength adjustor.

****

SAQ 3.2g

Describe two types of connection at the inner surface of a solid state ise. How do they function as inner reference solutions?

Response

The two common connections use either an internal solution or an all solid state configuration as shown in Fig. 3.2g:

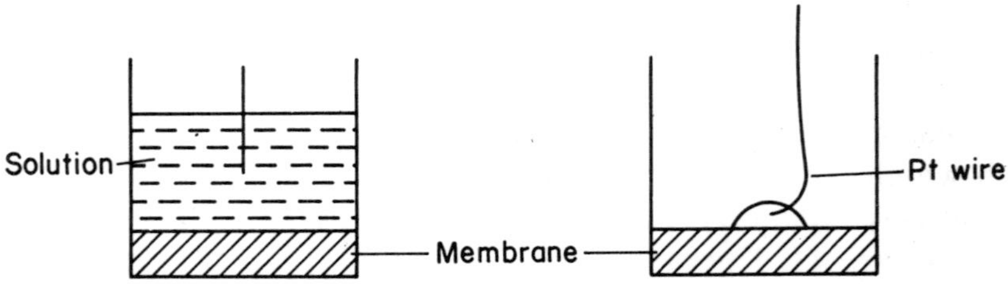

Fig. 3.2g. *Connections used at inner surface of a solid state ise*

With the internal solution of Ag^+ and a membrane of AgCl, an equilibrium is established:

$$Ag^+(\text{surface}) \rightleftharpoons Ag^+(\text{aq})$$

As the activity of $Ag^+(\text{aq})$ is a constant value, the equilibrium does not change. This maintains a steady boundary potential.

With the all solid state configuration, the inner equilibrium is:

$$Ag^+(\text{surface}) \rightleftharpoons Ag^+(\text{solid})$$

As the activity of the Ag(solid) is also a constant, this establishes a steady boundary potential.

****

SAQ 3.2h

Lead chromate is less soluble than lead sulphide, yet the latter is used with a silver salt to prepare a disc for a Pb^+ ise. Why is this the case?

Response

Firstly, there is no reason why a mixed $PbCrO_4/Ag_2CrO_4$ membrane will not work as a good selective membrane for Pb^{2+}.

The reason for using a mixed PbS/Ag_2S membrane lies in the ease with which the two salts can be co-precipitated from solution. It is a relatively simple matter to pass H_2S gas through a solution at an appropriate pH. The preparation of the mixed chromate membrane is not so easy.

****

SAQ 3.2i

Suggest how you could construct a solid state ise which would respond to CN^- ions in solution.

Response

The obvious choice here is to go for a membrane of AgCN. Unfortunately, AgCN does not crystallise out into three dimensional ionic lattices, but into a linear macromolecular structure of repeating AgCN units. It, therefore, does not have the lattice sites to produce the surface equilibria required, nor does it have a defect mechanism for the transfer of charge.

We can however make use of the fact that AgCN is less soluble than most silver halides. Consequently, in the absence of a halide in solution, any silver halide will act as a CN^- ise.

We generally use AgI as the membrane because it is the least soluble of the silver halides. The surface reaction is:

$$AgI(surface) + CN^-(aq) = AgCN(surface) + I^-(aq)$$

We are effectively using the interfering reaction to use the electrode as a CN^- ise. The two problems with this are:

(*i*) the surface must be cleaned regularly;

(*ii*) the membrane life is reduced to about 2 to 3 months.

****

SAQ 3.3a

Sketch a liquid membrane ion selective electrode, that makes use of an ion exchange compound.

Response

Your sketch should resemble Fig. 3.3a. You could also have drawn one of the newer screw-on modules. The basic idea of both is the same with a ion exchange compound dissolved in a water-immiscible organic solvent and absorbed into a filter.

****

SAQ 3.3b

List a few points that are important when selecting materials for ion exchange and neutral carrier ion selective electrodes.

Response

There are a few points you need to consider in selecting the active material:

(*i*) the material must form a strong complex with the ion under test;

(*ii*) the material must not form stable complexes with any other ion that is likely to be present. It is possible that one active material may be used for one application and another for a second application where different interfering ions are present;

(*iii*) the material must not be soluble in the test solution – generally this means that it must not be water-soluble;

(*iv*) the material must be non-toxic and not react with the test solution.

It is also necessary to select a solvent that dissolves the active material and is also immiscible with the test solution. Different solvents will affect the stability of the complexes and active material/solvent are selected in matched pairs.

A third variable that can be considered is whether to use a solid or liquid membrane.

****

SAQ 3.3c

Which of the following compounds could be used as an active material for a Cl^- ise?

(*i*) tetra-octylammonium nitrate;

(*ii*) dioctadecylphosphinous chloride;

(*iii*) chlorovaleric acid;

(*iv*) dimethyl-dioctadecylammonium chloride.

Response

The best compound to select from this list is dimethyl-dicotadecyl-ammonium chloride.

(*i*) is a poor choice as it does not contain free Cl^- ions, nor a grouping that will exchange with Cl^-.

(*ii*) looks a possibility on the surface. However, the structure is:

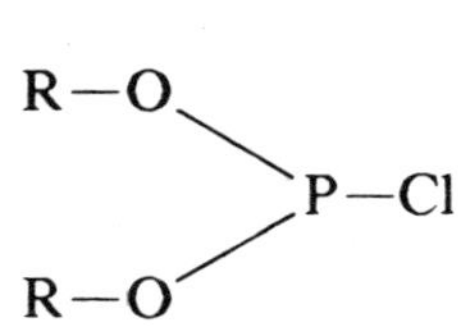

and the chloride is covalently bonded to the phosphorus. There are, therefore, no free Cl^- ions to set up the boundary potential.

(*iii*) as with the previous example, the chlorine is again covalently bonded within the molecule and there are no free Cl^- ions.

(*iv*) this is the best answer as the compound contains the free Cl^- ions necessary to form the boundary potential. It also contains a large organic grouping, which reduces its solubility in water.

****

SAQ 3.3d

Which of the following ions is likely to interfere with the operation of a potassium ise based on valinomycin?

Na^+	$k_{K,Na}$	=	10^{-4}
NH_4^+	k_{K,NH_4}	=	10^{-2}
Cs^+	$k_{K,Cs}$	=	10^{-1}
Rb^+	$k_{K,Rb}$	=	5
Li^+	$k_{K,Li}$	=	10^{-4}
Ca^{2+}	$k_{K,Ca}$	=	10^{-5}
Mg^{2+}	$k_{K,Mg}$	=	10^{-5}

Response

Whilst the values of selectivity coefficient will not give us an exact value for the interference, they do give us some idea as to where the problems can occur.

The two most likely sources of interference are Cs^+ and Rb^+. Fortunately, these two ions are not commonly found in large concentrations in most sample solutions and they cause a few problems.

The next most likely source of interference is the ammonium ion, and if this was present in large amounts in the test solution, it would have to be removed.

An ion, such as Na^+, which is commonly found in alkaline solutions, is not likely to cause a problem, unless its concentration is very high. Other ions with similarly small selectivity coefficients, are also unlikely to cause us too many problems unless all are present in the same solution – remember, the effect of interfering ions is cumulative.

In all cases, it must be stressed that the values of selectivity coefficient quoted for ion exchange and neutral carrier electrodes, should only be used as a guide.

****

SAQ 3.3e

Write a chemical equation to show how a caesium ion, Cs^+, can interfere with a potassium ion selective electrode using the neutral carrier valinomycin (V), which responds to a potassium ion, K^+.

Response

The interfering reaction can be represented by:

$$\underset{\text{(membrane surface)}}{K^+V} + Cs^+(aq) \rightleftharpoons \underset{\text{(membrane surface)}}{Cs^+V} + K^+(aq)$$

This reaction releases K^+ ions to the test solution and decreases the activity of the K^+ on the membrane surface. Valinomycin forms slightly more stable complexes with K^+ than Cs^+. This is reflected in the selectivity coefficient which is in the order of 10^{-1}.

****

SAQ 3.3f

Which of the following statements are true when applied to ion exchange and neutral carrier electrodes?

(*i*) The electrodes have very short response times.

(*ii*) The lower limit of measurement is comparable with that of solid-state membrane ion selective electrodes.

(*iii*) After coming into contact with a concentrated solution of an interfering ion, the electrode is prone to drift.

(*iv*) The presence of interfering ions increases the response time.

Response

(*i*) This is false – the surface equilibrium with ion exchange and neutral carrier electrodes takes longer to establish than with solid-state membrane electrodes. The response times with the former are much longer, typically between 15s and 30 min. Solid-state electrodes respond in one or two seconds.

(*ii*) This is also false. Electrodes based on ion exchange and neutral carriers have lower limits of measurement which are a function of the solubility of the active material and its solvent in the test solution. This gives a lower limit in the region of 10^{-6} mol dm^{-3}. Solid state membrane electrodes have lower limits which are a function of the solubility of the membrane itself. These are generally below 10^{-7} mol dm^{-3}.

(*iii*) This statement is true. If the concentration of the interfering ion is high, it can diffuse into the bulk of the membrane. This can diffuse outwards at a later stage causing drift.

(*iv*) This statement is true of all ion selective electrodes. In the first instance, interfering ions react with the active material on the surface of the membrane. This forms a surface layer on the electrode surface, which reduces response time.

****

SAQ 3.4a

Sketch a CO_2 gas sensing probe. Write down the equation representing the equilibrium in the internal electrolyte.

Response

Your electrode should resemble the ammonia probe in Fig. 3.4a. The internal equilibrium will involve CO_2, H^+, HCO_3^- and H_2O.

$$CO_2 + H_2O \rightleftharpoons HCO_3^- + H^+$$

As with the ammonia probe, the activity of the HCO_3^- along with that of the H_2O is constant. The activity of H^+ is, therefore, a function of the dissolved CO_2 that diffuses through the membrane.

****

SAQ 3.4b

Which of the following (there may be more than one) would cause interference if it was present in the test solution when a gas sensing probe was used to measure the activity of dissolved CO_2?

(*i*) NaCl;

(*ii*) CH_3COOH;

(*iii*) SO_2;

(*iv*) NaOH.

Response

The answer to this question is that both (*iii*) and (*iv*) would interfere with the determination.

Neither (*i*) nor (*ii*) would affect the electrodes response as they do not contain dissolved gases, nor do they react with the CO_2.

SO_2 would interfere with the probe because, as a gas, it would diffuse through the membrane and react with the aqueous internal electrolyte as follows:

$$SO_2 + H_2O \rightleftarrows HSO_3^- + H^+$$

As the internal ise is a glass electrode, it will clearly create a response and give a high CO_2 reading.

The NaOH will also interfere as it provides an alkaline solution which would neutralise the CO_2:

$$CO_2 + OH^- \rightleftarrows HCO_3^-$$

****

SAQ 3.4c

Why might an ammonia gas sensing probe drift for a long period when in use?

Response

The usual reason for drift is the transfer of water vapour across the membrane. This is particularly significant when the concentration of the internal electrolyte is very high.

****

SAQ 3.4d

What is the main reason for failure of gas sensing probes?

Response

Failure of gas sensing probes is generally due to failure of the membrane. This can occur in one of two ways. Either the pores of the membrane become blocked by deposits from the solution or the membrane fractures in use.

****

SAQ 3.5a

Suggest an internal ise for the following enzyme reaction:

$$\text{L-glutamine} + H_2O \xrightarrow{\text{glutaminase}} \text{L-glutamate} + NH_3$$

Response

The best electrode to choose for this reaction would be the ammonia gas sensing probe. This would detect the free NH_3 directly. An alternative would be to use a NH_4^+ electrode. To convert the NH_3 into NH_4^+, an acid layer would need to be introduced between the enzyme and the glass membrane. This can cause problems as enzymes are very sensitive to pH changes and the presence of an acid solution could inhibit their performance. This alternative, therefore, overcomplicates the determination.

****

SAQ 3.5b

> The glucose electrode (described in 3.5.1.) uses an I^- ise. Which of the following ions would interfere with the determination?
>
> Cl^-;
>
> Na^+;
>
> S^{2-};
>
> Ca^{2+}.

Response

The only ion which would cause significant interference is the S^{2-}.

The I^- ise uses a solid state AgI membrane and the only ions that will interefere are those forming less soluble salts with either component of the membrane. As NaI and CaI_2 are soluble, the two cations will cause us few problems. AgCl is certainly insoluble, but is more soluble than AgI. Ag_2S, on the other hand, is much less soluble than AgI and, consequently, the presence of S^{2-} would cause us many problems.

****

SAQ 3.5c

> List the factors which need to be controlled for accurate analysis with an enzyme ise.

Response

A number of factors need to be carefully controlled. The important ones are:

pH;

temperature;

response time;

interfering ions.

****

SAQ 3.6a

Which of the following factors is the main advantage of ion-selective field effect transistors?

(*i*) more effective membranes;

(*ii*) larger working range;

(*iii*) long lifetime;

(*iv*) small scale.

Response

The correct answer is (*iv*).

Ion selective field effect transistors use the same membranes as conventional ion – selective electrodes. They generally have a much smaller working range and very short lifetimes.

Their principal advantage is that they can be made very small and are ideal for biological and biochemical application.

****

SAQ 4.1a

> What is the purpose of adding cdta to the isa for the determination of F^-?

Response

The cdta is a complexing agent. It reacts with any mineral ions present, releasing any fluoride which may have been complexed:

$$MF^- \text{ (complex)} + \text{cdta} = \text{Mcdta(complex)} + F^-$$

****

SAQ 4.1b

> A student calibrated a fluoride ion-selective electrode with the following results

concentration/ppm	emf/mV
1.00	96.0
5.00	70.5
10.00	59.0
50.00	32.5
100.00	21.5

> Draw a calibration graph and estimate the concentration of samples which gave the following emf readings under the same condition: 73.0; 69.5; 42.0; and 63.5 mV.

Response

The graph Fig. 4.1a shows the calibration graph and the concentration of the four solutions are

4.2 ppm

5.2 ppm

28 ppm

7.4 ppm respectively

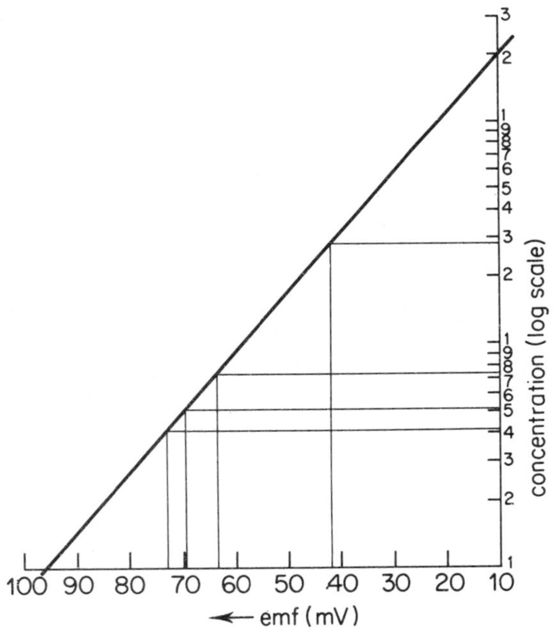

Fig. 4.1a. *Calibration curve for estimation of* F^-

****

SAQ 4.2a

> Why is the multiple standard addition method chosen for the analysis of Ca^{2+} in beer?

Response

Firstly, if we are only analysing one sample, there is little point in going to the effort of preparing several standards and plotting a calibration graph. Standard addition methods only use one standard solution and are quicker for one-off analyses.

Secondly, there is a high level of interfering species. The extrapolation procedure involved with a multiple standard addition method makes allowance for this. It would be very difficult to eliminate all the interfering species by pre-treatment.

****

SAQ 4.3a

> Why shouldn't a normal Ag,AgCl reference electrode be used with a nitrate ise?

Response

The reason is that the Cl^- ions interfere strongly with the NO_3^- ise. We could use the Ag,AgCl reference with a double junction, but eventually, the Cl^- ions would work their way through. A safer option is to use a mercury(I) sulphate reference.

****

SAQ 4.4a

Which of the following factors would be advantageous in selecting an ise for continuous monitoring?

(*i*) long lifetime;

(*ii*) low coefficient of thermal stability;

(*iii*) optimum pH operating range is identical to that of the sample;

(*iv*) smooth membrane surface.

Response

(*i*) This is certainly an important factor. Electrode lifetime is much shorter in continuous monitoring than in intermittent analysis. Not only is the actual cost of the replacement ise important, the time and effort needed to carry out the replacement is also significant.

(*ii*) This is not of great importance in this application as the temperatures of the samples and standards are controlled by thermostat.

(*iii*) Continuous monitors have the facility for adjusting the pH of the sample prior to measurement. Therefore, the operating range of the electrode is not important unless the pH adjustment would lead to precipitation of another species. If the ise dipped directly into the flow, then the pH range would be an important consideration.

(*iv*) In a flow system, a smooth membrane surface is favoured. If the membrane is rough, it can create turbulence which affects the flow over the surface. This, in turn, can lead to increased response times.

****

SAQ 4.5a

> If the number of samples of L-tyrosine per day were reduced to one or two, what steps would you change in the analysis?

Response

The electrode itself would remain the same. However, with only one or two samples, the use of a standard addition method in place of the calibration plot should be considered.

****

Units of Measurement

For historic reasons a number of different units of measurement have evolved to express quantity of the same thing. In the 1960s, many international scientific bodies recommended the standardisation of names and symbols and the adoption universally of a coherent set of units—the SI units (Système Internationale d'Unités)—based on the definition of five basic units: metre (m); kilogram (kg); second (s); ampere (A); mole (mol); and candela (cd).

The earlier literature references and some of the older text books, naturally use the older units. Even now many practicing scientists have not adopted the SI unit as their working unit. It is therefore necessary to know of the older units and be able to interconvert with SI units.

In this series of texts SI units are used as standard practice. However in areas of activity where their use has not become general practice, eg biologically based laboratories, the earlier defined units are used. This is explained in the study guide to each unit.

Table 1 shows some symbols and abbreviations commonly used in analytical chemistry; Table 5 is a glossary of abbreviations used in this particular text. Table 2 shows some of the alternative methods for expressing the values of physical quantities and the relationship to the value in SI units.

More details and definition of other units may be found in the *Manual of Symbols and Terminology for Physicochemical Quantities and Units*, Whiffen, 1979, Pergamon Press.

Table 1 *Symbols and Abbreviations Commonly used in Analytical Chemistry*

Å	Angstrom
$A_r(X)$	relative atomic mass of X
A	ampere
E or U	energy
G	Gibbs free energy (function)
H	enthalpy
J	joule
K	kelvin (273.15 + t °C)
K	equilibrium constant (with subscripts p, c, therm etc.)
K_a, K_b	acid and base ionisation constants
$M_r(X)$	relative molecular mass of X
N	newton (SI unit of force)
P	total pressure
s	standard deviation
T	temperature/K
V	volume
V	volt (J A^{-1} s^{-1})
a, a(A)	activity, activity of A
c	concentration/ mol dm^{-3}
e	electron
g	gramme
i	current
s	second
t	temperature / °C
bp	boiling point
fp	freezing point
mp	melting point
$\approx$	approximately equal to
$<$	less than
$>$	greater than
e, exp(x)	exponential of x
ln x	natural logarithm of x; ln x = 2.303 log x
log x	common logarithm of x to base 10

Table 2 *Alternative Methods of Expressing Various Physical Quantities*

1. **Mass (SI unit : kg)**

$$g = 10^{-3}\ kg$$
$$mg = 10^{-3}\ g = 10^{-6}\ kg$$
$$\mu g = 10^{-6}\ g = 10^{-9}\ kg$$

2. **Length (SI unit : m)**

$$cm = 10^{-2}\ m$$
$$Å = 10^{-10}\ m$$
$$nm = 10^{-9}\ m = 10Å$$
$$pm = 10^{-12}\ m = 10^{-2}\ Å$$

3. **Volume (SI unit : m^3)**

$$l = dm^3 = 10^{-3}\ m^3$$
$$ml = cm^3 = 10^{-6}\ m^3$$
$$\mu l = 10^{-3}\ cm^3$$

4. **Concentration (SI units : mol m^{-3})**

$$M = mol\ l^{-1} = mol\ dm^{-3} = 10^3\ mol\ m^{-3}$$
$$mg\ l^{-1} = \mu g\ cm^{-3} = ppm = 10^{-3}\ g\ dm^{-3}$$
$$\mu g\ g^{-1} = ppm = 10^{-6}\ g\ g^{-1}$$
$$ng\ cm^{-3} = 10^{-6}\ g\ dm^{-3}$$
$$ng\ dm^{-3} = pg\ cm^{-3}$$
$$pg\ g^{-1} = ppb = 10^{-12}\ g\ g^{-1}$$
$$mg\% = 10^{-2}\ g\ dm^{-3}$$
$$\mu g\% = 10^{-5}\ g\ dm^{-3}$$

5. **Pressure (SI unit : N m^{-2} = kg m^{-1} s^{-2})**

$$Pa = Nm^{-2}$$
$$atmos = 101\ 325\ N\ m^{-2}$$
$$bar = 10^5\ N\ m^{-2}$$
$$torr = mmHg = 133.322\ N\ m^{-2}$$

6. **Energy (SI unit : J = kg m^2 s^{-2})**

$$cal = 4.184\ J$$
$$erg = 10^{-7}\ J$$
$$eV = 1.602 \times 10^{-19}\ J$$

Table 3 *Prefixes for SI Units*

Fraction	Prefix	Symbol
10^{-1}	deci	d
10^{-2}	centi	c
10^{-3}	milli	m
10^{-6}	micro	μ
10^{-9}	nano	n
10^{-12}	pico	p
10^{-15}	femto	f
10^{-18}	atto	a

Multiple	Prefix	Symbol
10	deka	da
10^{2}	hecto	h
10^{3}	kilo	k
10^{6}	mega	M
10^{9}	giga	G
10^{12}	tera	T
10^{15}	peta	P
10^{18}	exa	E

Table 4 *Recommended Values of Physical Constants*

Physical constant	Symbol	Value
acceleration due to gravity	g	9.81 m s^{-2}
Avogadro constant	N_A	$6.022\,05 \times 10^{23} \text{ mol}^{-1}$
Boltzmann constant	k	$1.380\,66 \times 10^{-23} \text{ J K}^{-1}$
charge to mass ratio	e/m	$1.758\,796 \times 10^{11} \text{ C kg}^{-1}$
electronic charge	e	$1.602\,19 \times 10^{-19} \text{ C}$
Faraday constant	F	$9.648\,46 \times 10^{4} \text{ C mol}^{-1}$
gas constant	R	$8.314 \text{ J K}^{-1} \text{ mol}^{-1}$
'ice-point' temperature	T_{ice}	273.150 K exactly
molar volume of ideal gas (stp)	V_m	$2.241\,38 \times 10^{-2} \text{ m}^3 \text{ mol}^{-1}$
permittivity of a vacuum	ϵ_0	$8.854\,188 \times 10^{-12} \text{ kg}^{-1} \text{ m}^{-3} \text{ s}^4 \text{ A}^2 \text{ (F m}^{-1})$
Planck constant	h	$6.626\,2 \times 10^{-34} \text{ J s}$
standard atmosphere pressure	p	$101\,325 \text{ N m}^{-2}$ exactly
atomic mass unit	m_u	$1.660\,566 \times 10^{-27} \text{ kg}$
speed of light in a vacuum	c	$2.997\,925 \times 10^{8} \text{ m s}^{-1}$

Table 5 *Glossary and Abbreviations used in Electrochemistry*

E	emf
$E(X^+,X)$	electrode potential of X^+,X
$E^{\ominus}(X^+,X)$	standard electrode potential of X
E_j	liquid junction potential
F	Faraday constant
R	gas constant
S	Nernstian slope, 2.303 RT/F
T	temperature, K
a	activity
c	concentration
n	number of electrons transferred
λ	ionic activity coefficient

Other abbreviations

ABS	Acrylonitrile-butadiene-styrene copolymer
sce	Saturated Calomel Electrode
sep	Standard Electrode Potential
she	Standard Hydrogen Electrode
isa	Ionic Strength Adjustor
ise	Ion Selective Electrode
isfet	Ion Selective Field Effect Transistor
tisab	Total Ion Strength Adjustor Buffer